# STAAR Math Practice Grade 3

## Complete Content Review Plus 2 Full-length STAAR Math Tests

Elise Baniam - Michael Smith

**STAAR Math Practice Grade 3**
**Published in the United State of America By**
**The Math Notion**
**Email:** info@Mathnotion.com
**Web:** WWW.MathNotion.com

Copyright © 2020 by the Math Notion. All rights reserved. No part of this publication may be reproduced, stored in a retrieval system, or transmitted in any form or by any means, electronic, mechanical, photocopying, recording, scanning, or otherwise, except as permitted under Section 107 or 108 of the 1976 United States Copyright Ac, without permission of the author.
All inquiries should be addressed to the Math Notion.

**ISBN:** 978-1-63620-010-1

## About the Author

**Elise Baniam** has been a math instructor for over a decade now. She graduated in Mathematics. Since 2006, Elise has devoted his time to both teaching and developing exceptional math learning materials. As a Math instructor and test prep expert, Elise has worked with thousands of students. She has used the feedback of her students to develop a unique study program that can be used by students to drastically improve their math score fast and effectively.

- **– SAT Math Workbook**
- **– ACT Math Workbook**
- **– ISEE Math Workbooks**
- **– SSAT Math Workbooks**
- **–many Math Education Workbooks**
- **– and some Mathematics books ...**

As an experienced Math teacher, Mrs. Baniam employs a variety of formats to help students achieve their goals: she teaches students in large groups, and she provides training materials and textbooks through her website and through Amazon.

You can contact Elise via email at:
Elise@Mathnotion.com

# STAAR Math Practice Grade 3

**Get the Targeted Practice You Need to Excel on the Math Section of the STAAR Test Grade 3!**

**STAAR Math Practice Book** Grade 3 is **an excellent investment in your future** and the best solution for students who want to maximize their score and minimize study time. Practice is an essential part of preparing for a test and improving a test taker's chance of success. The best way to practice taking a test is by going through lots of STAAR math questions.

High-quality mathematics instruction ensures that students become problem solvers. We believe all students can develop deep conceptual understanding and procedural fluency in mathematics. In doing so, through this math workbook we help our students grapple with real problems, think mathematically, and create solutions.

**STAAR Math Practice Book** allows you to:

- Reinforce your strengths and improve your weaknesses
- Practice **2500+ realistic** STAAR math practice questions
- Exercise math problems in a variety of formats that provide intensive practice
- Review and study **Two Full-length STAAR Practice Tests** with detailed explanations

...and much more!

This Comprehensive STAAR Math Practice Book is carefully designed to provide only that **clear and concise information** you need.

# WWW.MathNotion.com

… So Much More Online!

✓ FREE Math Lessons

✓ More Math Learning Books!

✓ Mathematics Worksheets

✓ Online Math Tutors

**For a PDF Version of This Book**

**Please Visit WWW.MathNotion.com**

# Contents

**Chapter 1: Place Value and Number Sense** .................................................. 11
    Numbers in Standard Form .............................................................................. 12
    Number in Expand Form .................................................................................. 13
    Odd or Even ........................................................................................................ 14
    Compare Whole Numbers ................................................................................ 15
    Ordering Numbers ............................................................................................ 16
    Round Whole Numbers .................................................................................... 17
    Answer key Chapter 1 ...................................................................................... 18

**Chapter 2: Adding and Subtracting** .................................................................. 21
    Adding 3–Digit Numbers .................................................................................. 22
    Adding 4–Digit Numbers .................................................................................. 23
    Estimate Sums ................................................................................................... 24
    Subtracting 3–Digit Numbers .......................................................................... 25
    Subtracting 4–Digit Numbers .......................................................................... 26
    Estimate Differences ........................................................................................ 27
    Subtract from Whole Thousands. ................................................................... 28
    Answer key Chapter 2 ...................................................................................... 29

**Chapter 3: Multiplication and Division** ........................................................... 31
    Multiplication by 0 to 3 .................................................................................... 32
    Multiplication by 4 to 7 .................................................................................... 33
    Multiplication by 8 to 12 .................................................................................. 34
    Multiply Two Digit Number ............................................................................. 35
    Multiply Tens and Hundreds. .......................................................................... 36
    Division by 0 to 6 .............................................................................................. 37
    Division by 7 to 12 ............................................................................................ 38
    Dividing by Tens ............................................................................................... 39
    Advanced Division ........................................................................................... 40
    Times Table ....................................................................................................... 41
    Answers of Worksheets – Chapter 3 .............................................................. 43

**Chapter 4: Patterns** ............................................................................................ 47
    Repeating Pattern ............................................................................................. 48

Growing Patterns.................................................................................49
　　Patterns: Numbers.............................................................................50
　　Find a Rule........................................................................................51
　　Algebraic Thinking............................................................................52
　　Answers of Worksheets – Chapter 4..................................................53

**Chapter 5: Fractions and Mix Numbers............................................55**
　　Visual Fractions.................................................................................56
　　Adding Fractions................................................................................57
　　Subtracting Fractions.........................................................................58
　　Converting Mix Numbers..................................................................59
　　Simplify Fractions..............................................................................60
　　Addition Mix Numbers......................................................................61
　　Subtracting Mix Numbers..................................................................62
　　Comparing Fractions or Mix Numbers..............................................63
　　Answer key Chapter 5........................................................................64

**Chapter 6: Time and Money................................................................67**
　　Time....................................................................................................68
　　Time Conversion................................................................................69
　　Time Duration....................................................................................69
　　Money Amounts.................................................................................70
　　Money: Word Problems.....................................................................71
　　Answers of Worksheets – Chapter 6..................................................72

**Chapter 7: Measurement......................................................................73**
　　Reference Measurement.....................................................................74
　　Metric Length Measurement..............................................................75
　　Customary Length Measurement.......................................................75
　　Metric Capacity Measurement...........................................................76
　　Customary Capacity Measurement....................................................76
　　Metric Weight and Mass Measurement.............................................77
　　Customary Weight and Mass Measurement......................................77
　　Answers of Worksheets – Chapter 7..................................................78

**Chapter 8: Symmetry and Transformations......................................81**
　　Line Segments....................................................................................82
　　Parallel, Perpendicular and Intersecting Lines..................................83

Identify Lines of Symmetry ................................................................................. 84
Lines of Symmetry ............................................................................................... 85
Identify Three–Dimensional Figures .................................................................... 86
Answers of Worksheets – Chapter 8 .................................................................... 87

**Chapter 9: Geometry** ............................................................................................. **89**
Identifying Angles ................................................................................................ 90
Polygon Names .................................................................................................... 91
Triangles .............................................................................................................. 92
Quadrilaterals and Rectangles ............................................................................. 93
Area and Perimeter of Square ............................................................................. 94
Area and Perimeter of Rectangle ........................................................................ 95
Area and Perimeter of Triangle ........................................................................... 96
Perimeter of Polygon ........................................................................................... 97
Answer key Chapter 9 .......................................................................................... 99

**Chapter 10: Data and Graphs** ............................................................................... **101**
Tally and Pictographs ........................................................................................ 102
Dot plots ............................................................................................................ 103
Bar Graph .......................................................................................................... 104
Line Graphs ....................................................................................................... 105
Answer key Chapter 10 ...................................................................................... 106

**STAAR Test Review** ............................................................................................. **109**
STAAR Practice Test 1 ...................................................................................... 113
STAAR Practice Test 2 ...................................................................................... 127

**Answers and Explanations** ................................................................................. **139**
Answer Key ........................................................................................................ 141
Practice Test 1 ................................................................................................... 143
Practice Test 2 ................................................................................................... 148

# Chapter 1: Place Value and Number Sense

# Numbers in Standard Form

Write the number in standard form.

1) 10 million 208 thousand 24

    _____

2) 72 million 9 thousand 708

    _____

3) 121 million 24 thousand 453

    _____

4) 541 million 75 thousand 127

    _____

5) 90 billion 15 million 68 thousand 15

    _____

6) 12 billion 120 million 5

    _____

7) 8 billion 114 million 88 thousand

    _____

8) 16 billion 28 thousand 785

    _____

9) 75 billion 159 thousand 324

    _____

10) 41 billion 3 million 8 thousand 25

    _____

11) 16 billion 129 thousand 989

    _____

12) 65 billion 220 million 6 thousand 2

    _____

13) 785 million 124 thousand 97

    _____

14) 33 billion 104 million 11 thousand 57

    _____

15) 95 billion 424 million

    _____

16) 27 billion 77 million 9 thousand 150

    _____

# Number in Expand Form

Write the number in expand form.

1) 956: _____.

2) 3,800: _____.

3) 52,457: _____.

4) 60,070: _____.

5) 409,389: _____.

6) 76,805: _____.

7) 745,321: _____.

8) 8,146: _____.

9) 19,037: _____.

10) 52,799: _____.

11) 5,125: _____.

12) 400,544: _____.

13) 600,700: _____.

14) 3,080,000: _____.

## Odd or Even

Write odd or even.

1) 19 _____

2) 81 _____

3) 456 _____

4) 852 _____

5) 953 _____

6) 183 _____

7) 987 _____

8) 540 _____

9) 777 _____

10) 544 _____

11) 33 _____

12) 4,458 _____

13) 15,159 _____

14) 9,357 _____

15) 3,000 _____

16) 14 _____

17) 257 _____

18) 660 _____

19) 45,789 _____

20) 15,300 _____

21) 452 _____

22) 49,459 _____

23) 84 _____

24) 7,700 _____

25) 6,451 _____

26) 985 _____

## Compare Whole Numbers

Compare, writing <, >, or = between the numbers.

1) 40,420 ☐ 41,004

2) 29,460 ☐ 29,640

3) 78,920 ☐ 87,290

4) 34,570 ☐ 33,750

5) 96,328 ☐ 96,238

6) 85,843 ☐ 85,840

7) 76,584 ☐ 76,854

8) 72,998 ☐ 72,989

9) 37,467 ☐ 37,567

10) 48,878 ☐ 49,878

11) 56,660 ☐ 65,660

12) 73,898 ☐ 69,899

13) 89,990 ☐ 98,110

14) 84,760 ☐ 84,670

15) 26,680 ☐ 26,860

16) 86,440 ☐ 86,440

17) 158,980 ☐ 158,890

18) 201,807 ☐ 201,807

19) 243,240 ☐ 243,420

20) 345,566 ☐ 354,655

21) 187,158 ☐ 196,001

22) 137,983 ☐ 137,895

23) 278,788 ☐ 249,988

24) 194,854 ☐ 194,845

25) 219,390 ☐ 291,110

26) 305,288 ☐ 299,999

27) 317,857 ☐ 371,857

28) 405,710 ☐ 405,170

# STAAR Math Practice Grade 3

## Ordering Numbers

Write the numbers in order from the smallest to the greatest.

1) 312,585    89, 550    128, 550    213,858

_____

2) 422,599    405,290    427,260    99,080

_____

3) 87,800    212, 970    157, 970    221,758

_____

4) 545,100    455,100    48, 900    459,200

_____

5) 88,720    880, 720    808,720    800,820

_____

6) 743,200    437, 400    347, 500    734,300

_____

7) 280,240    281, 290    820, 000    182,920

_____

8) 157,420    175, 320    147, 250    417,210

_____

# Round Whole Numbers

Round to the place of the underlined digit.

1) 7,467,589 ≈ _____

2) 546,125 ≈ _____

3) 9,187,208 ≈ _____

4) 15,685,807 ≈ _____

5) 5,454,676 ≈ _____

6) 3,588,975 ≈ _____

7) 8,368,519 ≈ _____

8) 27,754,769 ≈ _____

9) 42,654,411 ≈ _____

10) 7,621,879 ≈ _____

11) 19,788,987 ≈ _____

12) 4,286,850 ≈ _____

13) 9,273,778 ≈ _____

14) 6,484,684 ≈ _____

15) 5,157,628 ≈ _____

16) 8,667,885 ≈ _____

17) 3,567,980 ≈ _____

18) 8,369,432 ≈ _____

19) 24,256,880 ≈ _____

20) 5,229,758 ≈ _____

21) 6,987,422 ≈ _____

22) 4,877,391 ≈ _____

# Answer key Chapter 1

**Numbers in Standard Form**

1) 10,208,024
2) 72,009,708
3) 121,024,453
4) 541,075,127
9) 75,000,159,324
10) 41,003,008,025
11) 16,000,129,989
12) 65,220,006,002
5) 90,015,068,015
6) 12,120,000,005
7) 8,114,088,000
8) 16,000,028,785
13) 785,124,097
14) 33,104,011,057
15) 95,424,000,000
16) 27,077,009,150

**Numbers in Expand Form**

1) (9 × 100) + (5 × 10) + 6
2) (3 × 1,000) + (8 × 100)
3) (5 × 10,000) + (2 × 1,000) + (4 × 100) + (5 × 10) + 7
4) (6 × 10,000) + (7 × 10) + 0
5) (4 × 100,000) + (9 × 1,000) + (3 × 100) + (8 × 10) + 9
6) (7 × 10,000) + (6 × 1,000) + (8 × 100) + 5
7) (7 × 100,000) + (4 × 10,000) + (5 × 1,000) + (3 × 100) + (2 × 10) + 1
8) (8 × 1,000) + (1 × 100) + (4 × 10) + 6
9) (1 × 10,000) + (9 × 1,000) + (3 × 10) + 7
10) (5 × 10,000) + (2 × 1,000) + (7 × 100) + (9 × 10) + 9
11) (5 × 1,000) + (1 × 100) + (2 × 10) + 5
12) (4 × 100,000) + (5 × 100) + (4 × 10) + 4
13) (6 × 100,000) + (7 × 100)
14) (3 × 1,000,000) + (8 × 10,000)

**Odd or Even**

1) Odd
2) Odd
3) Even
4) Even
5) Odd
6) Odd
7) Odd
8) Even
9) Odd
10) Even
11) Odd
12) Even

# STAAR Math Practice Grade 3

13) Odd
14) Odd
15) Even
16) Even
17) Odd
18) Even
19) Odd
20) Even
21) Even
22) Odd
23) Even
24) Even
25) Odd
26) Odd

**Compare Whole Numbers**

1) <
2) <
3) <
4) >
5) >
6) >
7) <
8) >
9) <
10) <
11) <
12) >
13) <
14) >
15) <
16) =
17) >
18) =
19) <
20) <
21) <
22) >
23) >
24) >
25) <
26) >
27) <
28) >

**Ordering Numbers**

1) 89, 550   128, 550   213,858   312,585
2) 99,080   405,290   422,599   427,260
3) 87,800   157, 970   212, 970   221,758
4) 48, 900   455,100   459,200   545,100
5) 88,720   800,820   808,720   880, 720
6) 347, 500   437, 400   734,300   743,200
7) 182,920   280,240   281, 290   820, 000
8) 147, 250   157,420   175, 320   417,210

**Round whole number**

1) 7,470,000
2) 546,000
3) 9,187,000
4) 15,686,000
5) 5,455,000
6) 3,589,000
7) 8,368,500
8) 27,754,800
9) 42,654,400
10) 7,600,000
11) 19,789,000
12) 4,287,000
13) 9,270,000
14) 6,485,000
15) 5,157,600
16) 8,668,000
17) 3,568,000
18) 8,369,430
19) 24,257,000
20) 5,230,000
21) 6,990,000
22) 4,877,000

# Chapter 2: Adding and Subtracting

# Adding 3-Digit Numbers

Find each sum.

1) 526 + 236

2) 725 + 130

3) 425 + 153

4) 563 + 125

5) 453 + 230

6) 398 + 120

7) 689 + 456

8) 863 + 325

9) 965 + 865

10) 369 + 120

11) 187 + 125

12) 389 + 150

13) 469 + 156

14) 360 + 150

15) 689 + 263

16) 890 + 345

17) 720 + 215

18) 680 + 230

# Adding 4-Digit Numbers

Add.

1)  2,135
   + 5,236
   _____

2)  4,369
   + 1,356
   _____

3)  6,598
   + 2,325
   _____

4)  3,125
   +4,035
   _____

5)  4,135
   +2,194
   _____

6)  5,036
   +2,365
   _____

7)  3,236
   +2,369
   _____

8)  6,320
   +3,765
   _____

9)  3,890
   +3,567
   _____

Find the missing numbers.

10) 1,155 + __ = 1,469

11) 400 + 3,000 = __

12) 5,200 + __ = 7,300

13) 555 + __ = 1,886

14) __ + 920 = 1,550

15) __ + 2,670 = 4,230

16) 689,505 = 80,000 + 600,000 + 5 + _____ + 9,000

17) 750,678 = 50,000 + 700,000 + 8 + _____ + 600

18) 574,962 = 70,000 + 500,000 + 2 + _____ + 900 + 60

# Estimate Sums

Estimate the sum by rounding each added to the nearest ten.

1) 36 + 9 =

2) 29 + 46 =

3) 36 + 12 =

4) 37 + 38 =

5) 12 + 35 =

6) 38 + 13 =

7) 48 + 25 =

8) 36 + 77 =

9) 45 + 86 =

10) 62 + 58 =

11) 45 + 36 =

12) 52 + 18 =

13) 35 + 59 =

14) 38 + 65 =

15) 87 + 82 =

16) 18 + 69 =

17) 65 + 64 =

18) 33 + 26 =

19) 73 + 48 =

20) 35 + 64 =

21) 13 + 93 =

22) 63 + 52 =

23) 164 + 142 =

24) 54 + 77 =

## Subtracting 3-Digit Numbers

Find the difference.

1) 756 − 236

2) 693 − 130

3) 425 − 153

4) 365 − 125

5) 493 − 230

6) 398 − 120

7) 989 − 756

8) 863 − 325

9) 965 − 465

10) 369 − 120

11) 159 − 125

12) 789 − 450

13) 469 − 156

14) 960 − 250

15) 689 − 358

16) 890 − 345

17) 929 − 115

18) 999 − 130

# Subtracting 4–Digit Numbers

Subtract.

1) 3,130 − 1,134

2) 3,356 − 2,870

3) 5,986 − 2,678

4) 6,987 − 6,422

5) 5,362 − 3,331

6) 7,365 − 2,212

7) 8,356 − 5,712

8) 8,350 − 2,729

9) 6,117 − 1,216

Find the missing number.

10) 4,223 − __ = 2,320

11) 5,856 − __ = 4,245

12) 1,136 − 689 = __

13) 4,200 − __ = 2,450

14) 5,870 − 2,650 = __

15) 6,360 − 4,320 = __

16) 8,165 − _____ = 4,303

17) 5,060 − 1,867 = __

18) Bob had $3,486 invested in the stock market until he lost $2,198 on those investments. How much money does he have in the stock market now?

# Estimate Differences

Estimate the difference by rounding each number to the nearest ten.

1) 58 − 23 =

2) 34 − 24 =

3) 75 − 48 =

4) 43 − 24 =

5) 69 − 46 =

6) 42 − 23 =

7) 77 − 47 =

8) 49 − 28 =

9) 94 − 48 =

10) 79 − 59 =

11) 68 − 26 =

12) 83 − 37 =

13) 73 − 43 =

14) 58 − 42 =

15) 82 − 52 =

16) 65 − 43 =

17) 99 − 81 =

18) 42 − 24 =

19) 58 − 47 =

20) 89 − 28 =

21) 81 − 65 =

22) 68 − 14 =

23) 76 − 6 =

24) 78 − 31 =

# Subtract from Whole Thousands.

Find the difference.

1) 3,000 − 10 = ___

2) 4,000 − 5 = ___

3) 2,000 − 8 = ___

4) 5,000 − 30 = ___

5) 7,000 − 7 = ___

6) 6,000 − 15 = ___

7) 8,000 − 40 = ___

8) 9,000 − 5 = ___

9) 2,000 − 8 = ___

10) 5,000 − 30 = ___

11) 7,000 − 200 = ___

12) 6,000 − 2 = ___

13) 4,000 − 20 = ___

14) 8,000 − 200 = ___

15) 5,000 − 100 = ___

16) 6,000 − 80 = ___

17) 5,000 − 70 = ___

18) 7,000 − 200 = ___

19) 9,000 − 300 = ___

20) 2,000 − 8 = ___

21) 4,000 − 10 = ___

22) 8,000 − 50 = ___

23) 3,000 − 90 = ___

24) 1,000 − 6 = ___

25) 5,000 − 5 = ___

26) 8,000 − 90 = ___

27) 9,000 − 30 = ___

28) 2,000 − 60 = ___

# Answer key Chapter 2

**Adding three–digit numbers**

1) 762
2) 855
3) 578
4) 688
5) 683
6) 518
7) 1,145
8) 1,188
9) 1,830
10) 489
11) 312
12) 539
13) 625
14) 510
15) 952
16) 1,235
17) 935
18) 910

**Adding 4–digit numbers**

1) 7,371
2) 5,725
3) 8,923
4) 7,160
5) 6,329
6) 7,401
7) 5,605
8) 10,085
9) 7,457
10) 314
11) 3,400
12) 2,100
13) 1,331
14) 630
15) 1,560
16) 500
17) 70
18) 4,000

**Estimate sums**

1) 50
2) 80
3) 50
4) 80
5) 50
6) 50
7) 80
8) 120
9) 140
10) 120
11) 90
12) 70
13) 100
14) 110
15) 170
16) 90
17) 130
18) 60
19) 120
20) 100
21) 100
22) 110
23) 300
24) 130

**Subtracting 3–digit numbers**

1) 520
2) 563
3) 272
4) 240
5) 263
6) 278
7) 233
8) 538
9) 500
10) 249
11) 34
12) 339
13) 313
14) 710
15) 331
16) 545
17) 814
18) 869

**Subtracting 4–digit numbers**

1) 1,996
2) 486
3) 3,308

4) 565
5) 2,031
6) 5,153
7) 2,644
8) 5,621
9) 4,901
10) 1,903
11) 1,611
12) 447
13) 1,750
14) 3,220
15) 2,040
16) 3,862
17) 3,193
18) 1,288

**Estimate differences**

1) 40
2) 10
3) 30
4) 20
5) 20
6) 20
7) 30
8) 20
9) 40
10) 20
11) 40
12) 40
13) 30
14) 20
15) 30
16) 30
17) 20
18) 20
19) 10
20) 60
21) 10
22) 60
23) 70
24) 50

**Subtract from Whole Thousands**

1) 2,990
2) 3,995
3) 1,992
4) 4,970
5) 6,993
6) 5,985
7) 7,960
8) 8,995
9) 1,992
10) 4,970
11) 6,800
12) 5,998
13) 3,980
14) 7,800
15) 4,900
16) 5,920
17) 5,930
18) 6,800
19) 8,700
20) 1,992
21) 3,990
22) 7,950
23) 2,910
24) 994
25) 4,995
26) 7,910
27) 8,970
28) 1,940

# Chapter 3: Multiplication and Division

## Multiplication by 0 to 3

Write the answers.

1) $5 \times 2 =$ ___

2) $5 \times 1 =$ ___

3) $7 \times 2 =$ ___

4) $7 \times 3 =$ ___

5) $1 \times 6 =$ ___

6) $2 \times 9 =$ ___

7) $10 \times 2 =$ ___

8) $1 \times 10 =$ ___

9) $11 \times 1 =$ ___

10) $5 \times 0 =$ ___

11) $9 \times 3 =$ ___

12) $11 \times 3 =$ ___

13) $2 \times 8 =$ ___

14) $3 \times 6 =$ ___

15) $3 \times 3 =$ ___

16) $8 \times 3 =$ ___

Find Each Missing Number.

17) $3 \times$ ___ $= 24$

18) $6 \times$ ___ $= 18$

19) $2 \times 4 =$ ___

20) ___ $\times 2 = 20$

21) ___ $\times 3 = 36$

22) $7 \times$ ___ $= 0$

23) $3 \times$ ___ $= 15$

24) $9 \times$ ___ $= 9$

25) $6 \times$ ___ $= 12$

26) ___ $\times 1 = 10$

27) ___ $\times 5 = 10$

28) $9 \times 0 =$ ___

29) $3 \times$ ___ $= 30$

30) ___ $\times 1 = 11$

# Multiplication by 4 to 7

Write the answers.

1) $7 \times 5 =$ ___

2) $5 \times 6 =$ ___

3) $10 \times 5 =$ ___

4) $12 \times 6 =$ ___

5) $9 \times 6 =$ ___

6) $11 \times 5 =$ ___

7) $8 \times 5 =$ ___

8) $7 \times 9 =$ ___

9) $20 \times 4 =$ ___

10) $4 \times 9 =$ ___

11) $11 \times 4 =$ ___

12) $9 \times 5 =$ ___

13) $8 \times 7 =$ ___

14) $12 \times 7 =$ ___

15) $8 \times 4 =$ ___

16) $5 \times 20 =$ ___

17) $7 \times 7 =$ ___

18) $6 \times 7 =$ ___

19) $4 \times 5 =$ ___

20) $4 \times 6 =$ ___

21) $4 \times 1 =$ ___

22) $7 \times 1 =$ ___

23) $6 \times 6 =$ ___

24) $4 \times 7 =$ ___

25) $10 \times 4 =$ ___

26) $12 \times 5 =$ ___

27) Ryan ordered six pizzas and sliced them into five pieces each. How many pieces of pizza were there? _____

## Multiplication by 8 to 12

Write the answers.

1) $8 \times 8 =$ ___

2) $10 \times 8 =$ ___

3) $11 \times 10 =$ ___

4) $9 \times 12 =$ ___

5) $10 \times 10 =$ ___

6) $12 \times 11 =$ ___

7) $11 \times 7 =$ ___

8) $9 \times 8 =$ ___

9) $7 \times 12 =$ ___

10) $11 \times 9 =$ ___

11) $10 \times 7 =$ ___

12) $3 \times 12 =$ ___

13) $8 \times 11 =$ ___

14) $12 \times 12 =$ ___

15) $8 \times 6 =$ ___

16) $12 \times 1 =$ ___

17) $12 \times 4 =$ ___

18) $10 \times 9 =$ ___

19) $11 \times 9 =$ ___

20) $12 \times 3 =$ ___

21) $11 \times 2 =$ ___

22) $9 \times 8 =$ ___

23) Each child has 8 books. If there are 10 children, how many books are there in total?

24) Each box has 10 pencils. If there are 13 boxes, how many pens are there in total?

# Multiply Two Digit Number

Find the answers.

1) 53 × 12

2) 46 × 10

3) 17 × 12

4) 45 × 14

5) 48 × 12

6) 45 × 21

7) 12 × 13

8) 42 × 20

9) 140 × 27

10) 564 × 24

11) 363 × 14

12) 36 × 20

13) 345 × 23

14) 725 × 30

15) 364 × 25

16) Emily has 17 candy bars. She divided each bar into 7 equal pieces to share with her colleagues. How many colleagues does Emily have? _____

17) Harper packaged cupcake in boxes of 12. She filled 36 boxes. How many cupcakes does Harper have? _____

## Multiply Tens and Hundreds.

Multiply, and find the missing factors.

1) 50 × 6 = _____

2) 7 × 400 = _____

3) 30 × 9 = _____

4) 80 × 70 = _____

5) 6 × 400 = _____

6) 70 × 90 = _____

7) 20 × 600 = _____

8) 10 × 900 = _____

9) 60 × 800 = _____

10) 7 × 700 = _____

11) _____ × 4 = 320

12) _____ × 8 = 6,400

13) _____ × 7 = 2,100

14) _____ × 9 = 720

15) _____ × 3 = 3,600

16) _____ × 600 = 5,400

17) _____ × 70 = 2,800

18) _____ × 30 = 2,700

19) _____ × 200 = 1,400

20) _____ × 300 = 12,000

21) 90 × _____ = 4,500

22) 40 × _____ = 2,400

23) 80 × _____ = 8,000

24) 60 × _____ = 420

25) 30 × _____ = 15,000

26) 700 × _____ = 6,3000

27) 50 × _____ = 3,000

28) 300 × _____ = 18,000

## Division by 0 to 6

Find each missing number.

1) 60 ÷ __ = 12

2) 32 ÷ 4 = __

3) 14 ÷ 2 = __

4) 35 ÷ 5 = __

5) __ ÷ 2 = 11

6) 10 ÷ 5 = __

7) __ ÷ 2 = 8

8) 15 ÷ __ = 5

9) __ ÷ 3 = 7

10) __ ÷ 4 = 6

11) __ ÷ 5 = 6

12) 19 ÷ 1 = __

13) 6 ÷ __ = 3

14) 22 ÷ 2 = __

15) 26 ÷ __ = 13

16) __ ÷ 4 = 9

17) 30 ÷ 6 = __

18) 28 ÷ __ = 7

19) 42 ÷ __ = 6

20) 12 ÷ __ = 3

21) 16 ÷ 2 = __

22) 10 ÷ 2 = __

23) __ ÷ 1 = 9

24) 27 ÷ __ = 9

25) 40 ÷ 4 = __

26) 11 ÷ __ = 1

27) 30 ÷ __ = 10

28) 34 ÷ __ = 17

29) 42 ÷ __ = 21

30) __ ÷ 5 = 12

31) Mia has 30 strawberries that she would like to give to her 3 friends. If she shares them equally, how many strawberries will she give to each of her friends?

# Division by 7 to 12

Find each missing number.

1) __ ÷ 13 = 1

2) __ ÷ 7 = 8

3) 90 ÷ 9 = __

4) 54 ÷ __ = 6

5) 81 ÷ 9 = __

6) __ ÷ 7 = 9

7) 42 ÷ __ = 6

8) __ ÷ 8 = 7

9) 36 ÷ __ = 4

10) 45 ÷ 9 = __

11) __ ÷ 10 = 10

12) 14 ÷ __ = 2

13) __ ÷ 12 = 9

14) 55 ÷ 11 = __

15) 110 ÷ __ = 10

16) 72 ÷ 8 = __

17) 96 ÷ 8 = __

18) 84 ÷ 7 = __

19) 80 ÷ __ = 10

20) 72 ÷ 12 = __

21) 36 ÷ __ = 3

22) 12 ÷ 12 = __

23) 147 ÷ __ = 21

24) 40 ÷ __ = 5

25) __ ÷ 8 = 11

26) 130 ÷ 10 = __

27) 121 ÷ __ = 11

28) 50 ÷ __ = 5

29) __ ÷ 12 = 2

30) __ ÷ 8 = 3

31) Stella has 64 fruit juice that she would like to give to her 8 friends. If she shares them equally, how many fruit juices will she give to each?

## Dividing by Tens

Find answers.

1) $600 \div 20 =$ ___

2) $300 \div 30 =$ ___

3) $1,400 \div 20 =$ ___

4) $450 \div 50 =$ ___

5) $320 \div 40 =$ ___

6) $100 \div 20 =$ ___

7) $360 \div 40 =$ ___

8) $210 \div 30 =$ ___

9) $1,500 \div 50 =$ ___

10) $220 \div 20 =$ ___

11) $770 \div 10 =$ ___

12) $180 \div 90 =$ ___

13) $260 \div 10 =$ ___

14) $540 \div 60 =$ ___

15) $160 \div 20 =$ ___

16) $510 \div 30 =$ ___

17) $200 \div 40 =$ ___

18) $400 \div 40 =$ ___

19) $\frac{420}{30} =$ ___

20) $\frac{800}{40} =$ ___

21) $\frac{700}{70} =$ ___

22) $\frac{660}{60} =$ ___

23) $\frac{280}{40} =$ ___

24) $\frac{350}{50} =$ ___

## Advanced Division

Find the quotient.

1) $5\overline{)100}=$

2) $8\overline{)64}=$

3) $13\overline{)169}=$

4) $3\overline{)24}=$

5) $12\overline{)144}=$

6) $8\overline{)48}=$

7) $2\overline{)12}=$

8) $7\overline{)21}=$

9) $9\overline{)108}=$

10) $5\overline{)30}=$

11) $4\overline{)36}=$

12) $13\overline{)65}=$

13) $8\overline{)56}=$

14) $9\overline{)90}=$

15) $11\overline{)121}=$

16) $5\overline{)90}=$

17) $2\overline{)36}=$

18) $8\overline{)24}=$

19) $4\overline{)60}=$

20) $9\overline{)153}=$

21) $6\overline{)114}=$

22) $5\overline{)90}=$

23) $10\overline{)170}=$

24) $11\overline{)132}=$

25) $4\overline{)540}=$

26) $8\overline{)640}=$

27) $8\overline{)216}=$

28) $8\overline{)112}=$

29) $9\overline{)495}=$

30) $20\overline{)400}=$

31) $11\overline{)484}=$

32) $10\overline{)800}=$

33) $2\overline{)64}=$

34) $3\overline{)48}=$

35) $4\overline{)76}=$

36) $12\overline{)720}=$

37) $8\overline{)160}=$

38) $6\overline{)750}=$

39) $9\overline{)378}=$

40) $4\overline{)812}=$

41) $5\overline{)1,025}=$

42) $3\overline{)489}=$

# Times Table

| ×  | 1  | 2  | 3  | 4  | 5  | 6  | 7  | 8  | 9  | 10  | 11  | 12  |
|----|----|----|----|----|----|----|----|----|----|-----|-----|-----|
| 1  | 1  | 2  | 3  | 4  | 5  | 6  | 7  | 8  | 9  | 10  | 11  | 12  |
| 2  | 2  | 4  | 6  | 8  | 10 | 12 | 14 | 16 | 18 | 20  | 22  | 24  |
| 3  | 3  | 6  | 9  | 12 | 15 | 18 | 21 | 24 | 27 | 30  | 33  | 36  |
| 4  | 4  | 8  | 12 | 16 | 20 | 24 | 28 | 32 | 36 | 40  | 44  | 48  |
| 5  | 5  | 10 | 15 | 20 | 25 | 30 | 35 | 40 | 45 | 50  | 55  | 60  |
| 6  | 6  | 12 | 18 | 24 | 30 | 36 | 42 | 48 | 54 | 60  | 66  | 72  |
| 7  | 7  | 14 | 21 | 28 | 35 | 42 | 49 | 56 | 63 | 70  | 77  | 84  |
| 8  | 8  | 16 | 24 | 32 | 40 | 48 | 56 | 64 | 72 | 80  | 88  | 96  |
| 9  | 9  | 18 | 27 | 36 | 45 | 54 | 63 | 72 | 81 | 90  | 99  | 108 |
| 10 | 10 | 20 | 30 | 40 | 50 | 60 | 70 | 80 | 90 | 100 | 110 | 120 |
| 11 | 11 | 22 | 33 | 44 | 55 | 66 | 77 | 88 | 99 | 110 | 121 | 132 |
| 12 | 12 | 24 | 36 | 48 | 60 | 72 | 84 | 96 | 108| 120 | 132 | 144 |

WWW.MathNotion.com

# Answers of Worksheets – Chapter 3

**Multiplication by 0 to 3**

1) 10
2) 5
3) 14
4) 21
5) 6
6) 18
7) 20
8) 10
9) 11
10) 0
11) 27
12) 33
13) 16
14) 18
15) 9
16) 24
17) 8
18) 3
19) 8
20) 10
21) 12
22) 0
23) 5
24) 1
25) 2
26) 10
27) 2
28) 0
29) 10
30) 11

**Multiplication by 4 to 7**

1) 35
2) 30
3) 50
4) 72
5) 54
6) 55
7) 40
8) 63
9) 80
10) 36
11) 44
12) 45
13) 56
14) 84
15) 32
16) 100
17) 49
18) 42
19) 20
20) 24
21) 4
22) 7
23) 36
24) 28
25) 40
26) 60
27) 30

**Multiplication by 8 to 12**

1) 64
2) 80
3) 110
4) 108
5) 100
6) 132
7) 77
8) 72
9) 84
10) 99
11) 70
12) 36
13) 88
14) 144
15) 48
16) 12
17) 48
18) 90
19) 99
20) 36
21) 22
22) 72
23) 80
24) 130

**Multiplication Two Digit Numbers**

1) 636
2) 460
3) 204
4) 630
5) 576
6) 945
7) 156
8) 840
9) 3,780
10) 13,536
11) 5,082
12) 720

# STAAR Math Practice Grade 3

13) 7,935
14) 21,750
15) 9,100
16) 119
17) 432

## Multiply Tens and Hundreds

1) 300
2) 2,800
3) 270
4) 560
5) 2,400
6) 6,300
7) 12,000
8) 9,000
9) 48,000
10) 4,900
11) 80
12) 800
13) 300
14) 80
15) 12,00
16) 9
17) 40
18) 90
19) 7
20) 40
21) 50
22) 60
23) 100
24) 7
25) 500
26) 90
27) 60
28) 60

## Division by 0 to 6

1) 4
2) 8
3) 7
4) 7
5) 22
6) 2
7) 16
8) 3
9) 21
10) 24
11) 30
12) 19
13) 2
14) 11
15) 2
16) 36
17) 5
18) 4
19) 7
20) 4
21) 8
22) 5
23) 9
24) 3
25) 10
26) 11
27) 3
28) 2
29) 2
30) 60
31) 10

## Division by 7 to 12

1) 13
2) 56
3) 10
4) 9
5) 9
6) 63
7) 7
8) 56
9) 9
10) 5
11) 100
12) 7
13) 108
14) 5
15) 11
16) 9
17) 12
18) 12
19) 8
20) 6
21) 12
22) 1
23) 7
24) 8
25) 88
26) 13
27) 11
28) 10
29) 24
30) 24
31) 8

# STAAR Math Practice Grade 3

**Dividing by tens**

1) 30
2) 10
3) 70
4) 9
5) 8
6) 5
7) 9
8) 7
9) 30
10) 11
11) 77
12) 2
13) 26
14) 9
15) 8
16) 17
17) 5
18) 10
19) 14
20) 20
21) 10
22) 11
23) 7
24) 7

**Advanced Division**

1) 20
2) 8
3) 13
4) 8
5) 12
6) 6
7) 6
8) 3
9) 12
10) 6
11) 9
12) 5
13) 7
14) 10
15) 11
16) 18
17) 18
18) 3
19) 15
20) 17
21) 19
22) 18
23) 17
24) 12
25) 135
26) 80
27) 27
28) 14
29) 55
30) 20
31) 44
32) 80
33) 32
34) 16
35) 19
36) 60
37) 20
38) 125
39) 42
40) 203
41) 205
42) 163

# STAAR Math Practice Grade 3

# Chapter 4: Patterns

# STAAR Math Practice Grade 3

## Repeating Pattern

Circle the picture that comes next in each picture pattern.

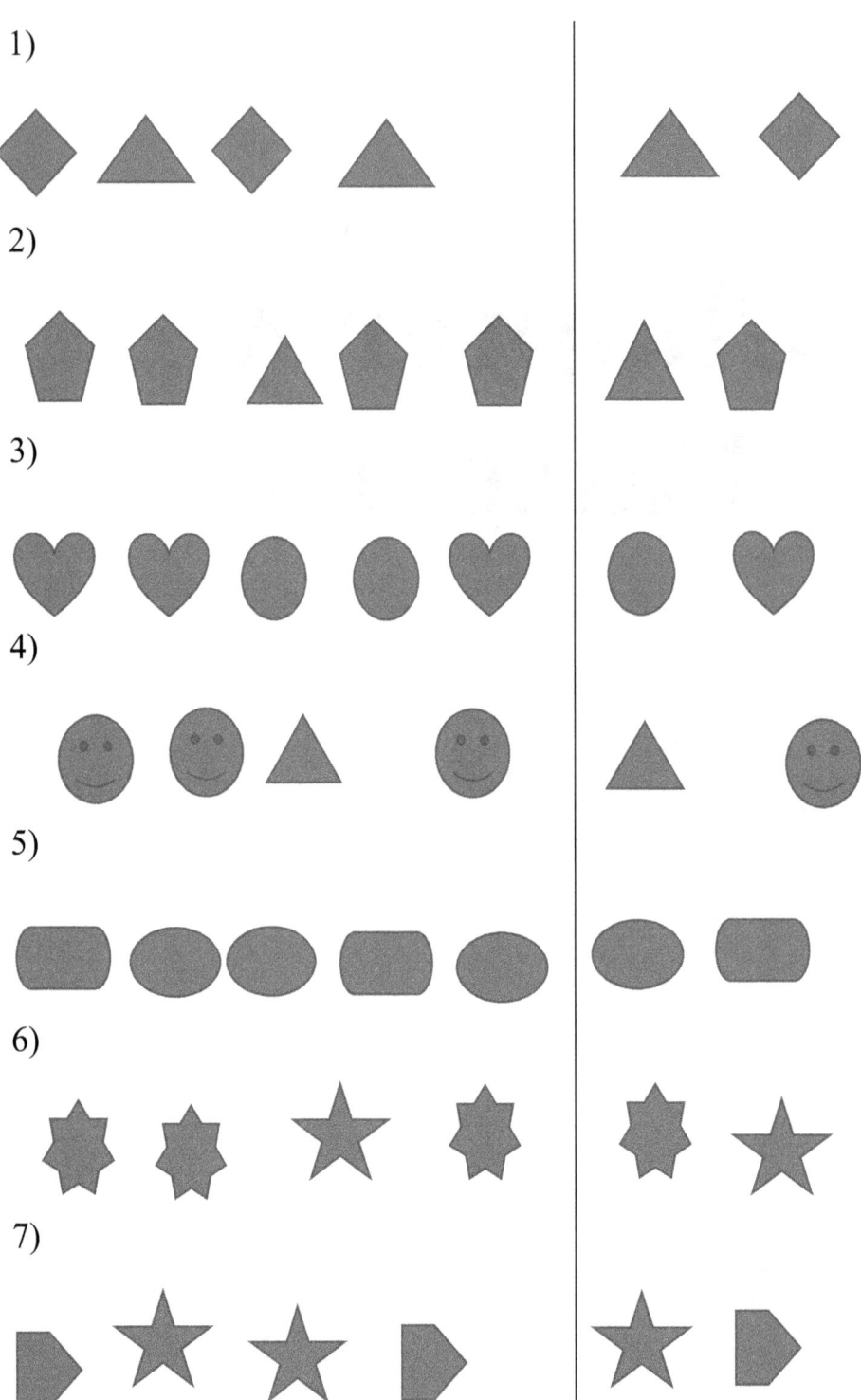

# Growing Patterns

Draw the picture that comes next in each growing pattern.

1)

2)

3)

4)

5)

6)

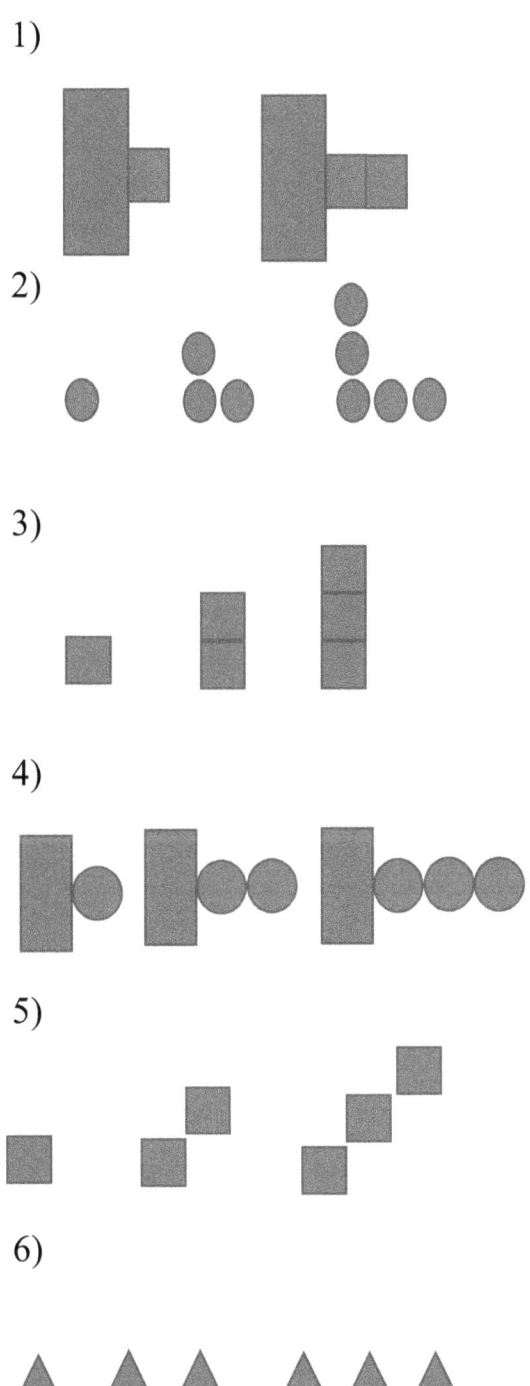

# Patterns: Numbers

Continue this pattern for four more numbers:

1) 1,500; 1,350; 1,200; 1,050; _____

2) 2,800; 2,600; 2,400; 2,200; _____

3) 3,500; 3,150; 2,800; 2,450; _____

4) 1,900; 1,780; 1,660; 1,540; _____

5) 3,200; 2,950; 2,700; 2,450; _____

6) 4,100; 3,800; 3,500; 3,200; _____

7) 5,400; 4,950; 4,500; 4,050; _____

8) 2,900; 2,725; 2,550; 2,375; _____

9) 1,950; 1,700; 1,450; 1,200; _____

10) 5,500; 4,900; 4,300; 3,700; _____

11) Write a list of five numbers that follows this pattern: Start at 100 and add 400 each time.

_____

# Find a Rule

Complete the output.

1- **Rule:** the output is $x + 25$

| Input | $x$ | 8 | 15 | 20 | 38 | 40 |
|---|---|---|---|---|---|---|
| Output | $y$ | | | | | |

2- **Rule:** the output is $x \times 18$

| Input | $x$ | 3 | 7 | 10 | 11 | 15 |
|---|---|---|---|---|---|---|
| Output | $y$ | | | | | |

3- **Rule:** the output is $x \div 7$

| Input | $x$ | 126 | 147 | 105 | 280 | 455 |
|---|---|---|---|---|---|---|
| Output | $y$ | | | | | |

Find a rule to write an expression.

4- **Rule:** _____

| Input | $x$ | 11 | 13 | 15 | 20 |
|---|---|---|---|---|---|
| Output | $y$ | 55 | 65 | 75 | 100 |

5- **Rule:** _____

| Input | $x$ | 10 | 28 | 32 | 46 |
|---|---|---|---|---|---|
| Output | $y$ | 14 | 32 | 36 | 50 |

6- **Rule:** _____

| Input | $x$ | 84 | 132 | 180 | 252 |
|---|---|---|---|---|---|
| Output | $y$ | 14 | 22 | 30 | 42 |

## Algebraic Thinking

Circle the number sentence that fits the problem. Then solve for x.

1) Mary had $42. Then she earned more money (x). Now she has $86.

   $42 + x = $86   OR   $42 + $86 = x

   x = ____

2) Lisa had $35. Then she earned more money (x). Now she has $78.

   $35 + x = $78   OR   $35 + $78 = x

   x = ____

3) Matthew had $37. Then he earned more money (x). Now he has $98.

   $37 + x = $98   OR   $37 + $98 = x

   x = ____

4) Charlotte gave 19 of the cookies he had baked to a friend and now he has 45 cookies left.   45 − 19 = x   OR   x − 19 = 45

   x = ____

5) Mia gave 32 of the cookies she had baked to a friend and now she has 55 cookies left.   55 − 32 = x   OR   x − 32 = 55

   x = ____

6) Lucas gave 41 of the cookies he had baked to a friend and now he has 49 cookies left. .   49 − 41 = x   OR   x − 41 = 49

   x = ____

# Answers of Worksheets – Chapter 4

**Repeating pattern**

1)

2)

3)

4)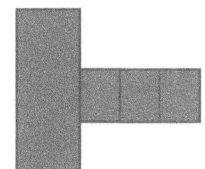

5)

6)

7)

**Growing patterns**

1)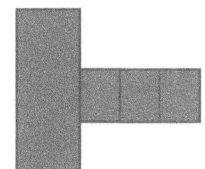

2)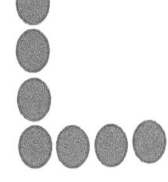

3)

4)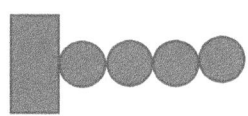

5)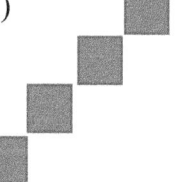

6)

**Pattern**

1) 900; 750; 600; 450

2) 2,000; 1,800; 1,600; 1,400

3) 2,100; 1,750; 1,400; 1,050

4) 1,420; 1,300; 1,180; 1,060

5) 2,200; 1,950; 1,700; 1,450

6) 2,900; 2,600; 2,300; 2,000

7) 3,600; 3,150; 2,700; 2,250

8) 2,200; 2,025; 1,850; 1,675

9) 950; 700; 450; 200

10) 3,100; 2,500; 1,900; 1,300

11) 100; 500; 900; 1,300; 1,700

**Find a Rule**

1)
| Input  | $x$ | 8  | 15 | 20 | 38 | 40 |
|--------|-----|----|----|----|----|----|
| Output | $y$ | 33 | 40 | 45 | 63 | **65** |

2)
| Input  | $x$ | 3  | 7   | 10  | 11  | 15  |
|--------|-----|----|-----|-----|-----|-----|
| Output | $y$ | 54 | **126** | 180 | 198 | 270 |

3)
| Input  | $x$ | 126 | 147 | 105 | 280 | 455 |
|--------|-----|-----|-----|-----|-----|-----|
| Output | $y$ | 18  | 21  | 15  | 40  | 65  |

4) y = 5x          5) y = x + 4          6) y = x ÷ 6

**Algebraic Thinking**

1) $42 + x = $86; x = 44

2) $35 + x = $78; x = 43

3) $37 + x = $98; x = 61

4) x − 19 = 45; x = 64

5) x − 32 = 55; x = 87

6) x − 41 = 49; x = 90

# Chapter 5: Fractions and Mix Numbers

# Visual Fractions

Write an addition to match the visual models.

1)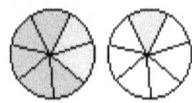

_____

2)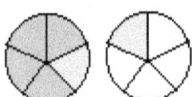

_____

3)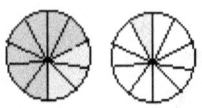

_____

4)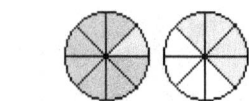

_____

5)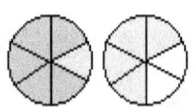

_____

6)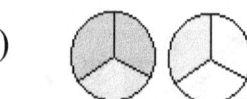

_____

7)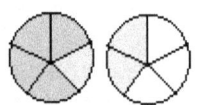

_____

8)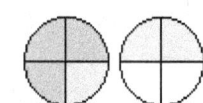

_____

9)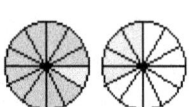

_____

10)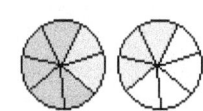

_____

## Adding Fractions

Find each sum.

1) $\frac{1}{4} + \frac{2}{4} =$

2) $\frac{2}{5} + \frac{1}{5} =$

3) $\frac{1}{8} + \frac{2}{8} =$

4) $\frac{4}{11} + \frac{1}{11} =$

5) $\frac{4}{21} + \frac{1}{21} =$

6) $\frac{5}{49} + \frac{6}{49} =$

7) $\frac{2}{7} + \frac{11}{7} =$

8) $\frac{1}{15} + \frac{3}{15} =$

9) $\frac{3}{19} + \frac{6}{19} =$

10) $\frac{1}{13} + \frac{1}{13} =$

11) $\frac{1}{5} + \frac{1}{5} =$

12) $\frac{4}{17} + \frac{6}{17} =$

13) $\frac{2}{20} + \frac{17}{20} =$

14) $\frac{4}{25} + \frac{7}{25} =$

15) $\frac{6}{14} + \frac{3}{14} =$

16) $\frac{12}{30} + \frac{5}{30} =$

17) $\frac{1}{9} + \frac{1}{9} =$

18) $\frac{29}{5} + \frac{3}{5} =$

19) $\frac{18}{6} + \frac{5}{6} =$

20) $\frac{25}{37} + \frac{11}{37} =$

## Subtracting Fractions

Find the difference.

1) $\dfrac{5}{3} - \dfrac{2}{3} =$

2) $\dfrac{5}{8} - \dfrac{3}{8} =$

3) $\dfrac{11}{14} - \dfrac{8}{14} =$

4) $\dfrac{13}{3} - \dfrac{7}{3} =$

5) $\dfrac{15}{17} - \dfrac{13}{17} =$

6) $\dfrac{18}{33} - \dfrac{10}{33} =$

7) $\dfrac{8}{25} - \dfrac{2}{25} =$

8) $\dfrac{17}{27} - \dfrac{2}{27} =$

9) $\dfrac{7}{10} - \dfrac{3}{10} =$

10) $\dfrac{24}{35} - \dfrac{4}{35} =$

11) $\dfrac{11}{5} - \dfrac{3}{5} =$

12) $\dfrac{28}{38} - \dfrac{18}{38} =$

13) $\dfrac{5}{6} - \dfrac{1}{6} =$

14) $\dfrac{22}{43} - \dfrac{11}{43} =$

15) $\dfrac{4}{7} - \dfrac{3}{7} =$

16) $\dfrac{18}{29} - \dfrac{15}{29} =$

17) $\dfrac{4}{5} - \dfrac{3}{5} =$

18) $\dfrac{42}{53} - \dfrac{38}{53} =$

19) $\dfrac{8}{31} - \dfrac{3}{31} =$

20) $\dfrac{32}{39} - \dfrac{30}{39} =$

21) $\dfrac{9}{26} - \dfrac{5}{26} =$

22) $\dfrac{31}{46} - \dfrac{27}{46} =$

23) $\dfrac{25}{48} - \dfrac{19}{48} =$

24) $\dfrac{39}{65} - \dfrac{27}{65} =$

# Converting Mix Numbers

Convert the following mixed numbers into improper fractions.

1) $3\frac{5}{6} =$

2) $5\frac{11}{15} =$

3) $4\frac{1}{3} =$

4) $2\frac{4}{7} =$

5) $7\frac{1}{4} =$

6) $3\frac{19}{21} =$

7) $5\frac{9}{10} =$

8) $4\frac{7}{12} =$

9) $3\frac{10}{11} =$

10) $6\frac{2}{5} =$

11) $8\frac{2}{3} =$

12) $2\frac{11}{12} =$

13) $3\frac{5}{6} =$

14) $4\frac{8}{11} =$

15) $7\frac{1}{4} =$

16) $5\frac{6}{11} =$

17) $8\frac{1}{5} =$

18) $3\frac{7}{12} =$

19) $6\frac{1}{22} =$

20) $3\frac{2}{3} =$

21) $7\frac{4}{5} =$

22) $4\frac{7}{8} =$

23) $6\frac{5}{6} =$

24) $12\frac{9}{10} =$

# Simplify Fractions

Reduce these fractions to lowest terms

1) $\frac{15}{10} =$

2) $\frac{20}{30} =$

3) $\frac{28}{35} =$

4) $\frac{21}{28} =$

5) $\frac{6}{18} =$

6) $\frac{27}{63} =$

7) $\frac{16}{28} =$

8) $\frac{48}{60} =$

9) $\frac{8}{72} =$

10) $\frac{30}{12} =$

11) $\frac{45}{60} =$

12) $\frac{30}{90} =$

13) $\frac{18}{30} =$

14) $\frac{5}{20} =$

15) $\frac{16}{56} =$

16) $\frac{56}{84} =$

17) $\frac{88}{33} =$

18) $\frac{36}{48} =$

19) $\frac{21}{56} =$

20) $\frac{64}{56} =$

21) $\frac{140}{280} =$

22) $\frac{30}{150} =$

23) $\frac{132}{11} =$

24) $\frac{13}{52} =$

## Addition Mix Numbers

Add the following fractions.

1) $1\frac{1}{5} + 4\frac{2}{5} =$

2) $5\frac{3}{7} + 3\frac{4}{7} =$

3) $2\frac{2}{8} + 3\frac{1}{8} =$

4) $5\frac{5}{8} + 3\frac{1}{8} =$

5) $2\frac{9}{12} + 3\frac{3}{12} =$

6) $6\frac{2}{3} + 3\frac{2}{3} =$

7) $2\frac{8}{27} + 2\frac{2}{27} =$

8) $2\frac{3}{4} + 3\frac{3}{4} =$

9) $4\frac{5}{6} + 1\frac{1}{6} =$

10) $3\frac{5}{7} + 1\frac{3}{7} =$

11) $12\frac{1}{2} + 2 =$

12) $5 + \frac{5}{6} =$

13) $5\frac{1}{3} + 2\frac{2}{3} =$

14) $3\frac{2}{12} + 3\frac{2}{12} =$

15) $4\frac{3}{5} + 4\frac{1}{5} =$

16) $5\frac{5}{7} + 1\frac{5}{7} =$

17) $4\frac{4}{6} + 6\frac{4}{6} =$

18) $2\frac{2}{5} + 3\frac{3}{5} =$

19) $3\frac{1}{9} + 2\frac{5}{9} =$

20) $5\frac{19}{25} + 3\frac{1}{25} =$

21) $4\frac{15}{43} + 1\frac{18}{43} =$

22) $6\frac{5}{9} + 4\frac{4}{9} =$

23) $2\frac{3}{35} + 3\frac{4}{35} =$

24) $3\frac{1}{42} + 1\frac{5}{42} =$

## STAAR Math Practice Grade 3

# Subtracting Mix Numbers

Subtract the following fractions.

1) $7\frac{1}{3} - 6\frac{1}{3} =$

2) $4\frac{5}{8} - 4\frac{2}{8} =$

3) $8\frac{5}{9} - 7\frac{1}{9} =$

4) $4\frac{1}{4} - 1\frac{1}{4} =$

5) $3\frac{1}{6} - 2\frac{5}{6} =$

6) $8\frac{1}{5} - 3\frac{2}{5} =$

7) $7\frac{5}{8} - 3\frac{3}{8} =$

8) $9\frac{9}{13} - 4\frac{6}{13} =$

9) $5\frac{7}{12} - 2\frac{5}{12} =$

10) $4\frac{4}{7} - 1\frac{3}{7} =$

11) $7\frac{1}{10} - 2\frac{7}{10} =$

12) $4\frac{5}{6} - 2\frac{1}{6} =$

13) $6\frac{2}{45} - \frac{17}{45} =$

14) $4\frac{13}{20} - 2\frac{13}{20} =$

15) $14\frac{4}{5} - 11\frac{2}{5} =$

16) $6\frac{2}{4} - 1\frac{1}{4} =$

17) $4\frac{1}{5} - 2\frac{3}{5} =$

18) $5\frac{5}{8} - 2\frac{1}{8} =$

19) $6\frac{45}{74} - 1\frac{45}{74} =$

20) $4\frac{3}{15} - 4\frac{1}{15} =$

21) $9\frac{9}{11} - 5\frac{1}{11} =$

22) $2\frac{41}{54} - 2\frac{29}{54} =$

23) $3\frac{25}{63} - 2\frac{4}{63} =$

24) $7\frac{9}{14} - 3\frac{3}{14} =$

WWW.MathNotion.com

# Comparing Fractions or Mix Numbers

Compare the fractions/mix numbers, and write >, < or =

1) $\frac{14}{3}$ _____ $\frac{24}{15}$

2) $\frac{32}{3}$ _____ $\frac{2}{5}$

3) $\frac{4}{9}$ _____ $\frac{2}{4}$

4) $\frac{12}{4}$ _____ $\frac{13}{9}$

5) $\frac{1}{8}$ _____ $\frac{2}{3}$

6) $\frac{10}{6}$ _____ $\frac{16}{7}$

7) $\frac{12}{13}$ _____ $\frac{7}{9}$

8) $\frac{20}{14}$ _____ $\frac{25}{3}$

9) $4\frac{1}{12}$ _____ $6\frac{1}{3}$

10) $8\frac{1}{6}$ _____ $3\frac{1}{8}$

11) $3\frac{1}{2}$ _____ $3\frac{1}{5}$

12) $7\frac{5}{8}$ _____ $7\frac{2}{9}$

13) $3\frac{2}{8}$ _____ $5\frac{3}{5}$

14) $\frac{1}{15}$ _____ $\frac{3}{7}$

15) $\frac{31}{25}$ _____ $\frac{19}{83}$

16) $\frac{12}{100}$ _____ $\frac{6}{62}$

17) $15\frac{1}{4}$ _____ $15\frac{1}{9}$

18) $\frac{1}{5}$ _____ $\frac{1}{9}$

19) $\frac{1}{7}$ _____ $\frac{1}{13}$

20) $\frac{1}{18}$ _____ $\frac{8}{15}$

21) $\frac{7}{22}$ _____ $\frac{9}{76}$

22) $\frac{4}{5}$ _____ $\frac{2}{5}$

23) $2\frac{17}{14}$ _____ $3\frac{3}{14}$

24) $3\frac{25}{4}$ _____ $4\frac{5}{4}$

# Answer key Chapter 5

**Visual Fractions**

1) $\frac{5}{7} + \frac{3}{7} = 1\frac{1}{7}$
2) $\frac{4}{5} + \frac{2}{5} = 1\frac{1}{5}$
3) $\frac{7}{10} + \frac{4}{10} = 1\frac{1}{10}$
4) $\frac{7}{8} + \frac{5}{8} = 1\frac{1}{2}$
5) $\frac{5}{6} + \frac{5}{6} = 1\frac{2}{3}$
6) $\frac{2}{3} + \frac{2}{3} = 1\frac{1}{3}$
7) $\frac{4}{5} + \frac{3}{5} = 1\frac{2}{5}$
8) $\frac{3}{4} + \frac{3}{4} = 1\frac{1}{2}$
9) $\frac{11}{12} + \frac{5}{12} = 1\frac{1}{3}$
10) $\frac{5}{7} + \frac{5}{7} = 1\frac{3}{7}$

**Adding Fractions**

1) $\frac{3}{4}$
2) $\frac{3}{5}$
3) $\frac{3}{8}$
4) $\frac{5}{11}$
5) $\frac{5}{21}$
6) $\frac{11}{49}$
7) $\frac{13}{7}$
8) $\frac{4}{15}$
9) $\frac{9}{19}$
10) $\frac{2}{13}$
11) $\frac{2}{5}$
12) $\frac{10}{17}$
13) $\frac{19}{20}$
14) $\frac{11}{25}$
15) $\frac{9}{14}$
16) $\frac{17}{30}$
17) $\frac{2}{9}$
18) $\frac{32}{5}$
19) $\frac{23}{6}$
20) $\frac{36}{37}$

**Subtracting Fractions**

1) $1$
2) $\frac{1}{4}$
3) $\frac{3}{14}$
4) $2$
5) $\frac{2}{17}$
6) $\frac{8}{33}$
7) $\frac{6}{25}$
8) $\frac{5}{9}$
9) $\frac{2}{5}$
10) $\frac{4}{7}$
11) $\frac{8}{5}$
12) $\frac{5}{19}$
13) $\frac{2}{3}$
14) $\frac{11}{43}$
15) $\frac{1}{7}$
16) $\frac{3}{29}$
17) $\frac{1}{5}$
18) $\frac{4}{53}$
19) $\frac{5}{31}$
20) $\frac{2}{39}$
21) $\frac{2}{13}$
22) $\frac{2}{23}$
23) $\frac{1}{8}$
24) $\frac{12}{65}$

# STAAR Math Practice Grade 3

**Converting Mix Numbers**

1) $\frac{23}{6}$
2) $\frac{86}{15}$
3) $\frac{13}{3}$
4) $\frac{18}{7}$
5) $\frac{29}{4}$
6) $\frac{82}{21}$
7) $\frac{59}{10}$
8) $\frac{55}{12}$
9) $\frac{43}{11}$
10) $\frac{32}{5}$
11) $\frac{26}{3}$
12) $\frac{35}{12}$
13) $\frac{23}{6}$
14) $\frac{52}{11}$
15) $\frac{29}{4}$
16) $\frac{61}{11}$
17) $\frac{41}{5}$
18) $\frac{43}{12}$
19) $\frac{133}{22}$
20) $\frac{11}{3}$
21) $\frac{39}{5}$
22) $\frac{39}{8}$
23) $\frac{41}{6}$
24) $\frac{129}{10}$

**Simplify Fractions**

1) $\frac{3}{2}$
2) $\frac{2}{3}$
3) $\frac{4}{5}$
4) $\frac{3}{4}$
5) $\frac{1}{3}$
6) $\frac{3}{7}$
7) $\frac{4}{7}$
8) $\frac{4}{5}$
9) $\frac{1}{9}$
10) $\frac{5}{2}$
11) $\frac{3}{4}$
12) $\frac{1}{3}$
13) $\frac{3}{5}$
14) $\frac{1}{4}$
15) $\frac{2}{7}$
16) $\frac{2}{3}$
17) $\frac{8}{3}$
18) $\frac{3}{4}$
19) $\frac{3}{8}$
20) $\frac{8}{7}$
21) $\frac{1}{2}$
22) $\frac{1}{5}$
23) 12
24) $\frac{1}{4}$

**Adding Mix Numbers**

1) $5\frac{3}{5}$
2) 9
3) $5\frac{3}{8}$
4) $8\frac{3}{4}$
5) 6
6) $10\frac{1}{3}$
7) $4\frac{10}{27}$
8) $6\frac{1}{2}$
9) 6
10) $5\frac{1}{7}$
11) $14\frac{1}{2}$
12) $5\frac{5}{6}$
13) 8
14) $6\frac{1}{3}$
15) $8\frac{4}{5}$

WWW.MathNotion.com

16) $7\frac{3}{7}$

17) $11\frac{1}{3}$

18) 6

19) $5\frac{2}{3}$

20) $8\frac{4}{5}$

21) $5\frac{33}{43}$

22) 11

23) $5\frac{1}{5}$

24) $4\frac{1}{7}$

**Subtracting Mix Numbers**

1) 1

2) $\frac{3}{8}$

3) $1\frac{4}{9}$

4) 3

5) $\frac{1}{3}$

6) $4\frac{4}{5}$

7) $4\frac{1}{4}$

8) $5\frac{3}{13}$

9) $3\frac{1}{6}$

10) $3\frac{1}{7}$

11) $4\frac{2}{5}$

12) $2\frac{2}{3}$

13) $5\frac{2}{3}$

14) 2

15) $3\frac{2}{5}$

16) $5\frac{1}{4}$

17) $1\frac{3}{5}$

18) $3\frac{1}{2}$

19) 5

20) $\frac{2}{15}$

21) $4\frac{8}{11}$

22) $\frac{2}{9}$

23) $1\frac{1}{3}$

24) $4\frac{3}{7}$

**Comparing Fractions**

1) >

2) >

3) <

4) >

5) <

6) <

7) >

8) <

9) <

10) >

11) >

12) >

13) <

14) <

15) >

16) <

17) >

18) >

19) >

20) <

21) >

22) >

23) =

24) >

# Chapter 6: Time and Money

**Unit of Time Conversion:**

- 1 year = 12 months
- 1 year = 52 weeks
- 1 week = 7 days
- 1 day = 24 hours
- 1 hour = 60 minutes
- 1 minute = 60 seconds

# Time

Write the time the clock shows, and the time 15 minutes late.

| 1) | a. ___:___ | b. ___:___ | c. ___:___ |
|---|---|---|---|
| 15 min. later | d. ___:___ | e. ___:___ | f. ___:___ |
| 2) | a. ___:___ | b. ___:___ | c. ___:___ |
| 15 min. later | d. ___:___ | e. ___:___ | f. ___:___ |
| 3) | a. ___:___ | b. ___:___ | c. ___:___ |
| 15 min. later | d. ___:___ | e. ___:___ | f. ___:___ |

# Time Conversion

Convert to the units.

1) 8 hr. = _____ min

2) 5 year = _____ week

3) 2 hr. = _____ sec

4) 8 min = _____ sec

5) 300 min = _____ hr.

6) 2 weeks = _____ min

7) 1 week = _____ hr.

8) 3 days = _____ hr

9) 1 day = _____ min

10) 480 min = _____ hr

11) 9 years = _____ month

12) 600 sec = _____ min

13) 48 hr. = _____ day

14) 15 weeks = _____ day

# Time Duration

How much time has passed?

1) From 3:35 A.M. to 6:45 A.M.: ____ hours and ___ minutes.

2) From 2:30 A.M. to 7:15 A.M.: ____ hours and ___ minutes.

3) It's 6:20 P.M. What time was 3 hours ago? _____ O'clock

4) 4:15 A.M to 7:35 AM: _____ hours and _____ minutes.

5) 1:45 A.M to 4:20 AM: _____ hours and _____ minutes.

6) 10:35 A.M. to 3:05 PM. = _____ hour(s) and _____ minutes

7) 5:12 A.M. to 5:48 A.M. = _____ minutes

8) 8:08 A.M. to 8:45 A.M. = _____ minutes

# Money Amounts

Add.

1) $314 + $132        $524 + $410       $390 + $215

2) $521 + $330        $630 + $321       $732 + $145

3) $511 + $212        $660 + $128       $830 + $110

4) $721.60 + $63.70   $221.20 + $220.75  $515.00 + $456.30

Subtract.

5) $836 − $155        $642 − $111       $733 − $533

6) $438 − $136        $498 − $326       $740 − $549

7) $356.40 − $219.70  $710.50 − $128.80  $832.70 − $379.20

8) Linda had $14.00. She bought some game tickets for $7.14. How much did she have left?

# Money: Word Problems

Solve.

1) How many books can you buy with $48 if one book costs $6?

2) After paying $8.15 for a sandwich, Ava has $44.36. How much money did she have before buying the salad?

3) How many packages of cream cheese can you buy with $90 if one package costs $6?

4) Last week Liam ran 30 miles more than Lucas. Liam ran 56 miles. How many miles did Lucas run?

5) Last Friday Logan had $32.52. Over the weekend he received some money for cleaning the attic. He now has $51. How much money did he receive?

6) After paying $12.12 for a pizza, Benjamin has $47.50. How much money did he have before buying the pizza?

# Answers of Worksheets – Chapter 6

**Time**

1) a. 11: 34;   b. 13: 27;   c. 15: 42
   d. 11: 49;   e. 3: 42;   f. 15: 57
2) a. 14: 38;   b. 08: 12;   c. 16: 57
   d. 14: 53;   e. 08: 27;   f. 17: 12
3) a. 12: 51;   b. 16: 38;   c. 11: 08
   d. 13: 06;   e. 16: 53;   f. 11: 23

**Time Conversion**

1) 480 min
2) 260 weeks
3) 7,200 sec
4) 480 sec
5) 5 hr
6) 20,160 min
7) 168 hr
8) 72 hr
9) 1,440 min
10) 8 hr
11) 108 months
12) 10 min
13) 2 days
14) 105 days

**Time Duration**

1) 3:10
2) 4:45
3) 3:20 P.M.
4) 3:20
5) 2:35
6) 4:30
7) 36 minutes
8) 37 minutes

**Add Money**

1) 446, 934, 605
2) 851, 951, 877
3) 723, 788, 940
4) 785.30, 441.95, 971.30

**Subtract Money**

5) 681–531–200
6) 302–172–191
7) 136.70–581.70–453.50
8) $6.86

**Money: word problem**

1) 8
2) $52.51
3) 15
4) 26
5) 18.48
6) 59.62

# Chapter 7: Measurement

# Reference Measurement

| LENGTH | |
|---|---|
| **Customary** | **Metric** |
| 1 mile (mi) = 1,760 yards (yd) | 1 kilometer (km) = 1,000 meters (m) |
| 1 yard (yd) = 3 feet (ft) | 1 meter (m) = 100 centimeters (cm) |
| 1 foot (ft) = 12 inches (in.) | 1 centimeter(cm) = 10 millimeters(mm) |
| **VOLUME AND CAPACITY** | |
| **Customary** | **Metric** |
| 1 gallon (gal) = 4 quarts (qt) | 1 liter (L) = 1,000 milliliters (mL) |
| 1 quart (qt) = 2 pints (pt.) | |
| 1 pint (pt.) = 2 cups (c) | |
| 1 cup (c) = 8 fluid ounces (Fl oz) | |
| **WEIGHT AND MASS** | |
| **Customary** | **Metric** |
| 1 ton (T) = 2,000 pounds (lb.) | 1 kilogram (kg) = 1,000 grams (g) |
| 1 pound (lb.) = 16 ounces (oz) | 1 gram (g) = 1,000 milligrams (mg) |
| **Time** | |
| 1 year = 12 months | |
| 1 year = 52 weeks | |
| 1 week = 7 days | |
| 1 day = 24 hours | |
| 1 hour = 60 minutes | |
| 1 minute = 60 seconds | |

# Metric Length Measurement

Convert to the units.

1) 2,000 mm = _____ cm

2) 5 m = _____ mm

3) 7 m = _____ cm

4) 9 km = _____ m

5) 5,000 mm = _____ m

6) 2,800 cm = _____ m

7) 13 m = _____ cm

8) 4,000 mm = _____ cm

9) 20,000 mm = _____ m

10) 7 km = _____ mm

11) 6 km = _____ m

12) 3 m = _____ cm

13) 17,000 m = _____ km

14) 500,000 m = _____ km

# Customary Length Measurement

Convert to the units.

1) 15 ft = _____ in

2) 8 ft = _____ in

3) 7 yd = _____ ft

4) 9 yd = _____ ft

5) 3 yd = _____ in

6) 3 mi = _____ in

7) 7,200 in = _____ yd

8) 252 in = _____ yd

9) 8,800 yd = _____ mi

10) 12 yd = _____ in

11) 4 mi = _____ yd

12) 47,520 ft = _____ mi

13) 60 in = _____ ft

14) 25 yd = _____ ft

15) 36 in = _____ ft

16) 2 mi = _____ ft

## Metric Capacity Measurement

Convert the following measurements.

1) 50 l = _____ ml

2) 4 l = _____ ml

3) 13 l = _____ ml

4) 8 l = _____ ml

5) 19 l = _____ ml

6) 2 l = _____ ml

7) 70,000 ml = _____ l

8) 8,000 ml = _____ l

9) 37,000 ml = _____ l

10) 200,000 ml = _____ l

11) 6,000,000 ml = _____ l

12) 40,000 ml = _____ l

## Customary Capacity Measurement

Convert the following measurements.

1) 2 gal = _____ qt.

2) 11 gal = _____ pt.

3) 3 gal = _____ c.

4) 14 pt. = _____ c

5) 43 c = _____ fl oz

6) 16 qt = _____ pt.

7) 8 qt = _____ c

8) 29 pt. = _____ c

9) 6,720 c = _____ gal

10) 144 pt. = _____ gal

11) 72 qt = _____ gal

12) 92 pt. = _____ qt

13) 4,600 c = _____ qt

14) 146 c = _____ pt.

15) 108 qt = _____ gal

16) 1,848 pt. = _____ qt

17) 31 gal = _____ pt.

18) 6 qt = _____ c

19) 640 c = _____ gal

20) 104 fl oz = _____ c

## Metric Weight and Mass Measurement

Convert.

1) 7 kg = _____ g

2) 3 kg = _____ g

3) 13 kg = _____ g

4) 21 kg = _____ g

5) 9 kg = _____ g

6) 121 kg = _____ g

7) 249 kg = _____ g

8) 4,000 g = _____ kg

9) 6,000 g = _____ kg

10) 17,000 g = _____ kg

11) 129,000 g = _____ kg

12) 220,000 g = _____ kg

13) 9,000,000 g = _____ kg

14) 11,000,000 g = _____ kg

## Customary Weight and Mass Measurement

Convert.

1) 16,000 lb. = _____ T

2) 20,000 lb. = _____ T

3) 170,000 lb. = _____ T

4) 44,000 lb. = _____ T

5) 7 lb. = _____ oz

6) 4 lb. = _____ oz

7) 10 lb. = _____ oz

8) 24 T = _____ lb.

9) 3 T = _____ lb.

10) 9 T = _____ lb.

11) 112 T = _____ lb.

12) 2 T = _____ oz

13) 5 T = _____ oz

14) 224 oz = _____ lb.

# Answers of Worksheets – Chapter 7

**Metric length**

1) 200 cm
2) 5,000 mm
3) 700 cm
4) 9,000 m
5) 5 m
6) 28 m
7) 1,300 cm
8) 400 cm
9) 20 m
10) 7,000,000 mm
11) 6,000 m
12) 300 cm
13) 17 km
14) 500 km

**Customary Length**

1) 180
2) 96
3) 21
4) 27
5) 108
6) 190,080
7) 200
8) 7
9) 5
10) 432
11) 7,040
12) 9
13) 5
14) 75
15) 3
16) 10,560

**Metric Capacity**

1) 50,000 ml
2) 4,000 ml
3) 13,000 ml
4) 8,000 ml
5) 19,000 ml
6) 2,000 ml
7) 70 L
8) 8 L
9) 37 L
10) 200 L
11) 6,000 L
12) 40 L

**Customary Capacity**

1) 8 qt
2) 88 pt.
3) 48 c
4) 28 c
5) 344 fl oz
6) 12 pt.
7) 32 c
8) 58 c
9) 420 gal
10) 18 gal
11) 18 gal
12) 46 qt
13) 1,150 qt
14) 73 pt.
15) 27 gal
16) 927 qt
17) 248 pt.
18) 24 c
19) 40 gal
20) 13 c

**Metric Weight and Mass**

1) 7,000 g
2) 3,000 g
3) 13,000 g
4) 21,000 g
5) 9,000 g
6) 121,000 g
7) 249,000 g
8) 4 kg
9) 6 kg
10) 17 kg
11) 129 kg
12) 220 kg
13) 9,000 kg
14) 11,000 kg

**Customary Weight and Mass**

1) 8 T

2) 10 T

3) 85 T

4) 22 T

5) 112 oz

6) 64 oz

7) 160 oz

8) 48,000 lb.

9) 6,000 lb.

10) 18,000 lb.

11) 224,000 lb.

12) 64,000 oz

13) 160,000 oz

14) 14 lb

# Chapter 8: Symmetry and Transformations

# Line Segments

Write each as a line, ray, or line segment.

1)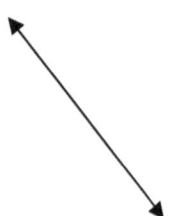

2)

3)

4)

5)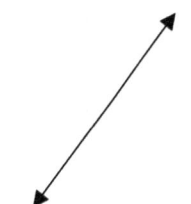

6)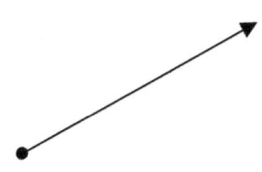

7)

8)

# Parallel, Perpendicular and Intersecting Lines

State whether the given pair of lines are parallel, perpendicular, or intersecting.

1)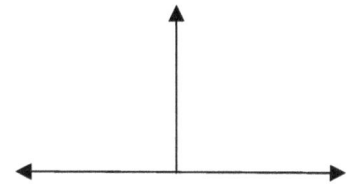

2)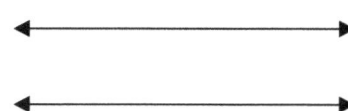

3)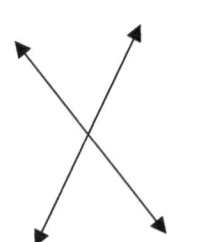

4)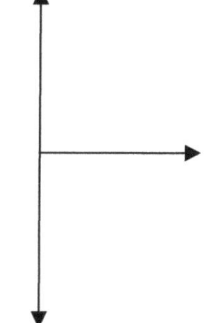

5)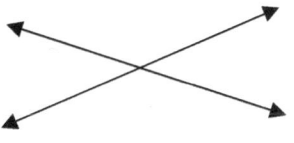

6)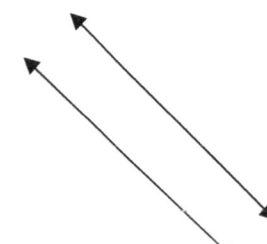

7)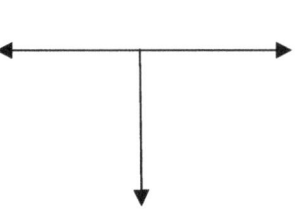

8)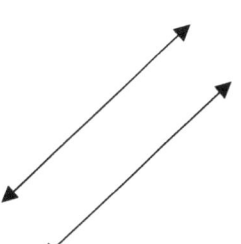

## Identify Lines of Symmetry

Tell whether the line on each shape a line of symmetry is.

1)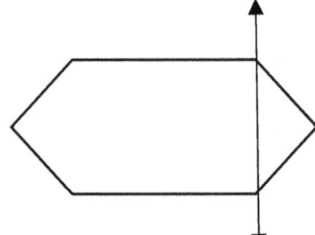

2)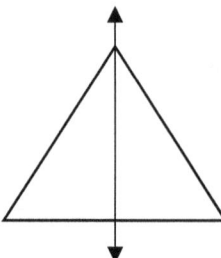

3)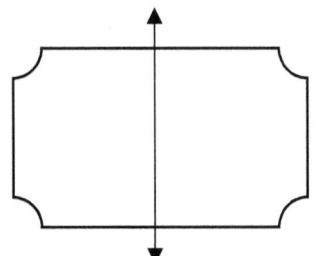

4)

5)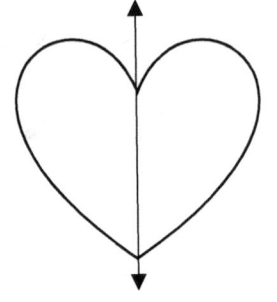

6)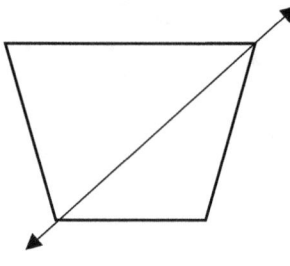

7)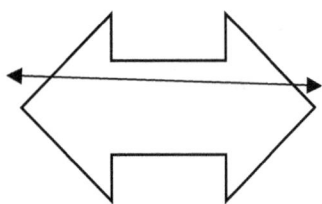

8)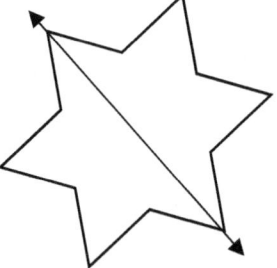

# Lines of Symmetry

Draw lines of symmetry on each shape. Count and write the lines of symmetry you see.

1)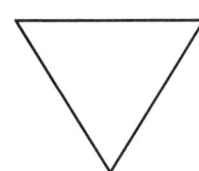

2)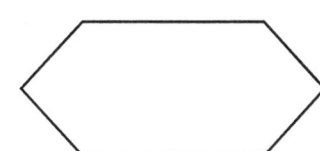

3)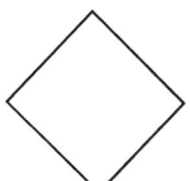

4)

5)

6)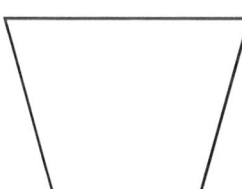

7)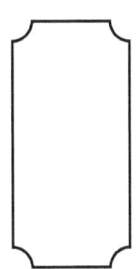

8)

## Identify Three–Dimensional Figures

Write the name of each shape.

1)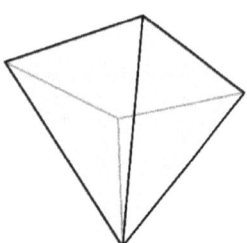

2)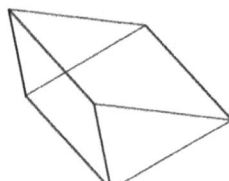

3)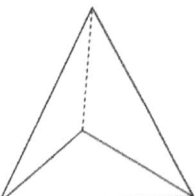

4)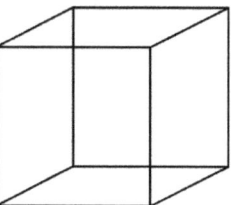

5)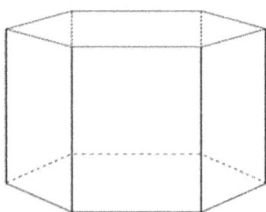

6)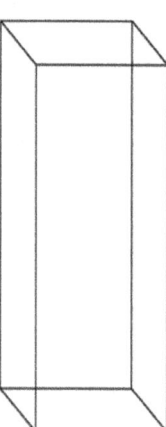

7)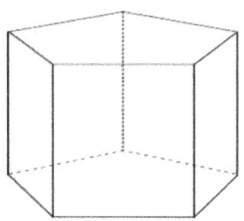

# Answers of Worksheets – Chapter 8

### Line Segments

1) Line
2) Line segment
3) Line segment
4) Ray
5) Line
6) Ray
7) Line segment
8) Ray

### Parallel, Perpendicular and Intersecting Lines

1) Perpendicular
2) Parallel
3) Intersection
4) Perpendicular
5) Intersection
6) Parallel
7) Perpendicular
8) Parallel

### Identify lines of symmetry

1) No
2) yes
3) yes
4) No
5) yes
6) No
7) No
8) yes

### lines of symmetry

1)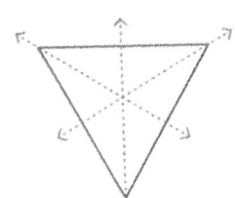

2)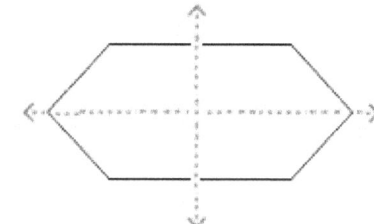

3)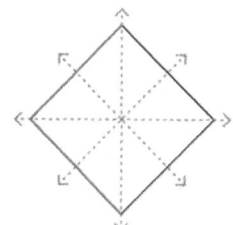

4)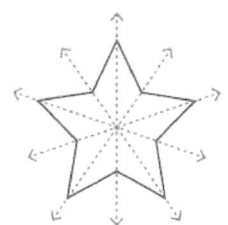

WWW.MathNotion.com

5)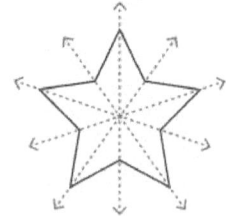

6)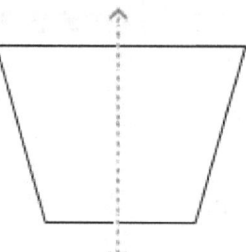

7)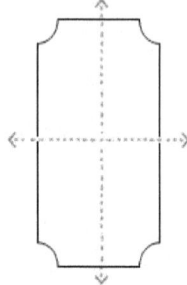

8)

**Identify Three–Dimensional Figures**

1) Square pyramid
2) Triangular prism
3) Triangular pyramid
4) Cube

5) Hexagonal prism
6) Rectangular prism
7) Pentagonal prism

# Chapter 9:
# Geometry

# Identifying Angles

Write the name of the angles( Acute, Right, Obtuse, and Straight Angles) .

1)

2)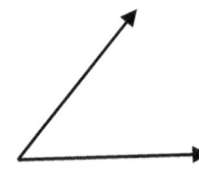

3)

4)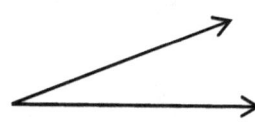

5)

6)

7)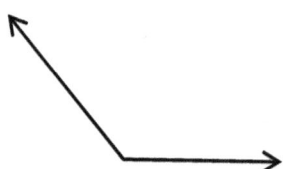

8)

# Polygon Names

Write name of polygons.

1)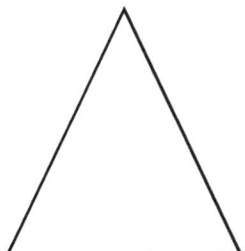

2)

3)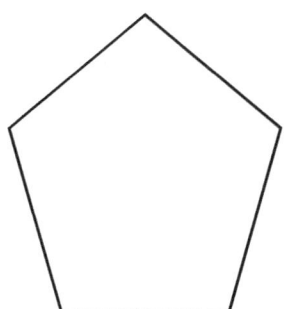

4)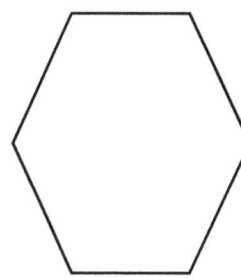

5)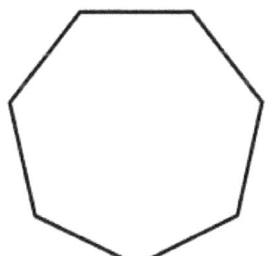

6)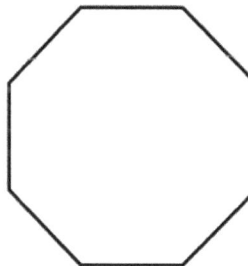

7)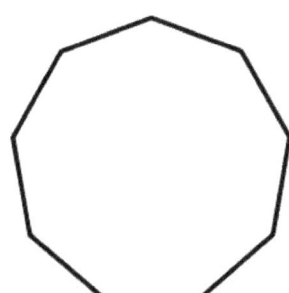

8)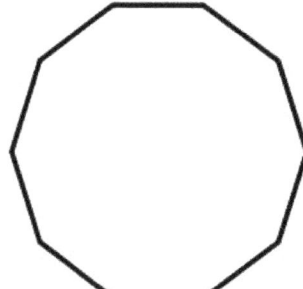

## STAAR Math Practice Grade 3

# Triangles

Classify the triangles by their sides and angles.

1)

2)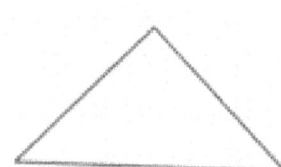

3)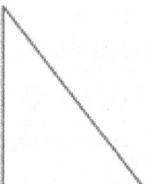

4)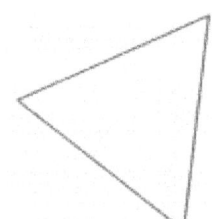

5)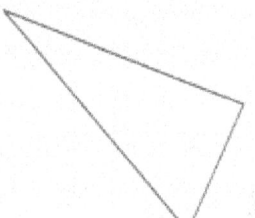

6)

Find the measure of the unknown angle in each triangle.

7)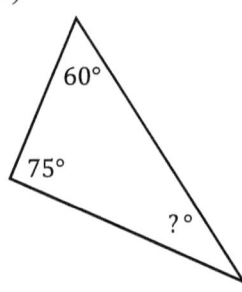

8)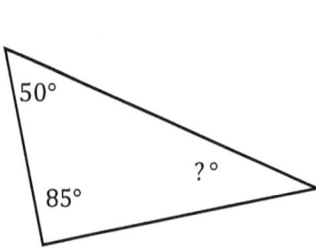

9)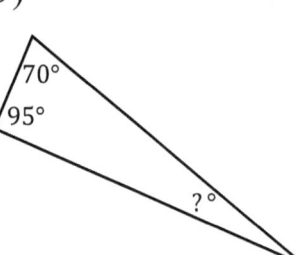

10)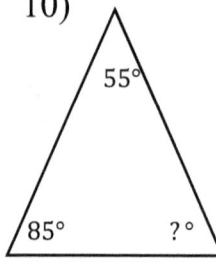

11)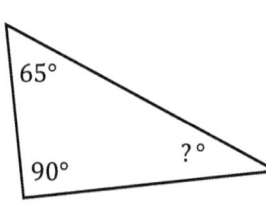

12)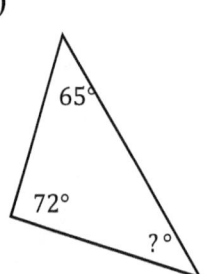

13)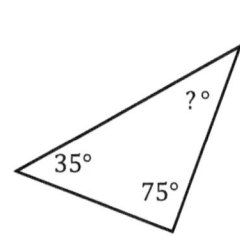

14)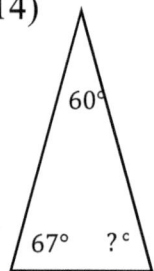

WWW.MathNotion.com

# Quadrilaterals and Rectangles

Write the name of quadrilaterals.

1)

2)

3)

4)

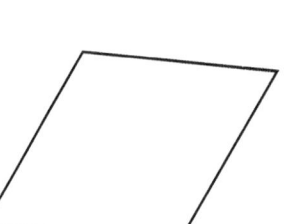

5)

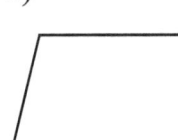

6)

Solve.

7) A rectangle has _____ sides and _____ angles.

8) Draw a rectangle that is 6 centimeters long and 5 centimeters wide. What is the perimeter?

9) Draw a rectangle 5 cm long and 3 cm wide.

10) Draw a rectangle whose length is 6cm and whose width is 4 cm. What is the perimeter of the rectangle?

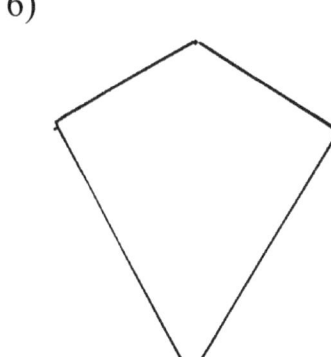

11) What is the perimeter of the rectangle?

# Area and Perimeter of Square

Find the perimeter and area of each squares.

1)

 4

Perimeter: ............:
Area: ............:

2)

 $1\frac{1}{2}$

Perimeter: ............:
Area: ............:

3)

2.5
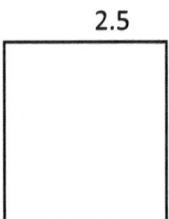

Perimeter: ............:
Area: ............:

4)

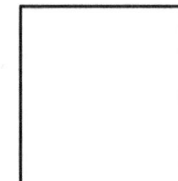 8

Perimeter: ............:
Area: ............:

5)

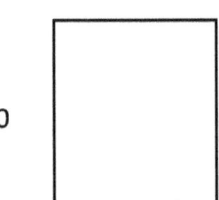
10

Perimeter: ............:
Area: ............:

6)

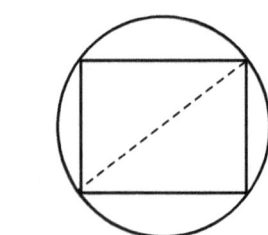
d=6

Perimeter of Square: ............:
Area of Square: ............:

# Area and Perimeter of Rectangle

Find the perimeter and area of each rectangle.

1)

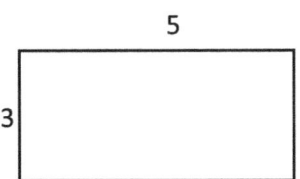

Perimeter: _____:

Area: _____:

2)

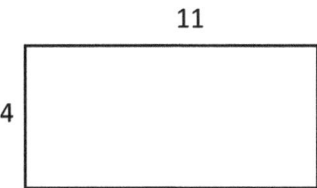

Perimeter: _____:

Area: _____:

3)

```
        10
   ┌─────────┐
1.2│         │
   └─────────┘
```

Perimeter: _____:

Area: _____:

4)

```
        9½
   ┌─────────┐
  2│         │
   └─────────┘
```

Perimeter: _____:

Area: _____:

5)

```
        7.2
   ┌─────────┐
  5│         │
   └─────────┘
```

Perimeter: _____:

Area: _____:

6)

```
        8.4
   ┌─────────┐
3.6│         │
   └─────────┘
```

Perimeter: _____:

Area: _____:

STAAR Math Practice Grade 3

# Area and Perimeter of Triangle

Find the perimeter and area of each triangle.

1)

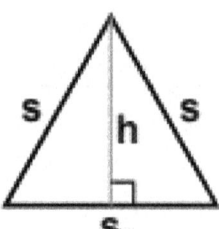

Perimeter: _____:

Area: _____:

2)

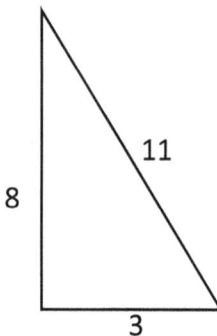

Perimeter: _____:

Area: _____:

3)

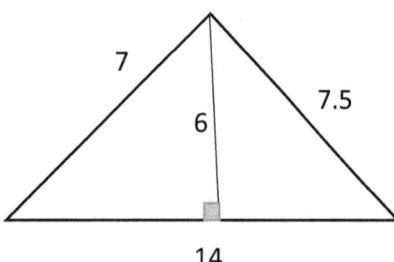

Perimeter: _____:

Area: _____.

4)

s=10

h=6.4

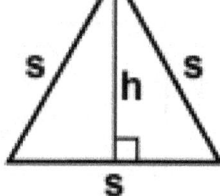

Perimeter: _____:

Area: _____:

5)

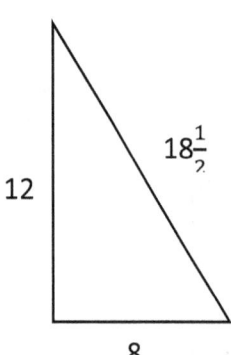

Perimeter: _____:

Area: _____:

6)

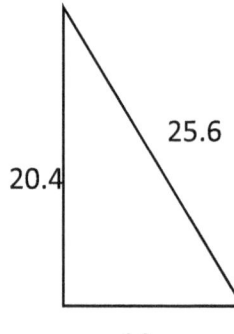

Perimeter: _____:

Area: _____:

WWW.MathNotion.com

# Perimeter of Polygon

Find the perimeter of each polygon.

1)

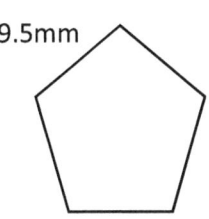
9.5mm

Perimeter: _____ :

2)

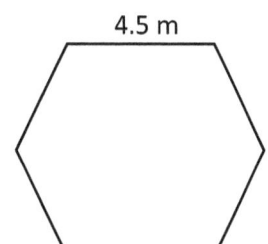
4.5 m

Perimeter: _____ :

3)

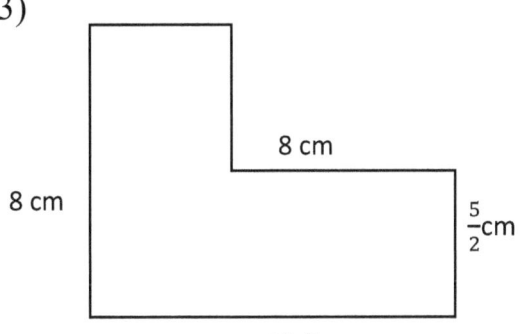
8 cm, 8 cm, $\frac{5}{2}$ cm, 12.5 cm

Perimeter: _____ :

4)

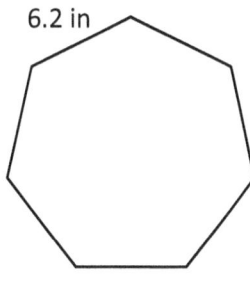
6.2 in

Perimeter: _____ :

5)

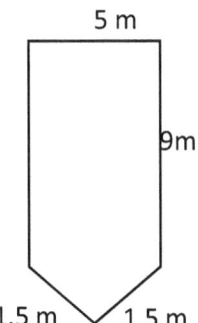
5 m, 9m, 1.5 m, 1.5 m

Perimeter: _____ :

6)

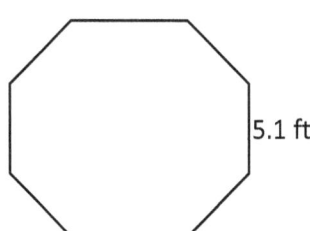
5.1 ft

Perimeter: _____ :

# Answer key Chapter 9

**Identifying Angles**

1) Obtuse
2) Acute
3) Right
4) Acute
5) Straight
6) Obtuse
7) Obtuse
8) Acute

**Polygon Names**

1) Triangle
2) Quadrilateral
3) Pentagon
4) Hexagon
5) Heptagon
6) Octagon
7) Nonagon
8) Decagon

**Triangles**

1) Scalene, obtuse
2) Isosceles, right
3) Scalene, right
4) Equilateral, acute
5) Scalene, acute
6) Scalene, acute
7) 45°
8) 45°
9) 15°
10) 40°
11) 25°
12) 43°
13) 70°
14) 53°

**Quadrilaterals and Rectangles**

1) Square
2) Rectangle
3) Parallelogram
4) Rhombus
5) Trapezoid
6) Kike
7) 4 - 4
8) 22
9) Use a rule to draw the rectangle
10) 20
11) 26

**Area and Perimeter of Square**

1. Perimeter: 16,   Area: 16
2. Perimeter: 6,   Area: 2.25
3. Perimeter: 10, Area: 6.25
4. Perimeter: 32,   Area: 64
5. Perimeter: 40,   Area: 100
6. Perimeter: $4\sqrt{3}$, Area: 3

**Area and Perimeter of Rectangle**

1- Perimeter: 16,   Area: 15
2- Perimeter: 30,   Area: 44
3- Perimeter: 22.4, Area: 12
4- Perimeter: 23, Area: 19
5- Perimeter: 24.4, Area: 36
6- Perimeter: 24, Area: 30.24

**Area and Perimeter of Triangle**

1- Perimeter: 3s, Area: $\frac{1}{2}sh$
2- Perimeter: 22,   Area: 12
3- Perimeter: 28.5, Area: 42
4- Perimeter: 30,   Area: 32
5- Perimeter: 38.5, Area: 48
6- Perimeter: 60, Area: 142.8

**Perimeter of Polygon**

1) 47.5 mm

2) 27 m

3) 41 cm

4) 43.4 in

5) 26 m

6) 40.8 ft

# Chapter 10: Data and Graphs

# Tally and Pictographs

Using the key, draw the pictograph to show the information.

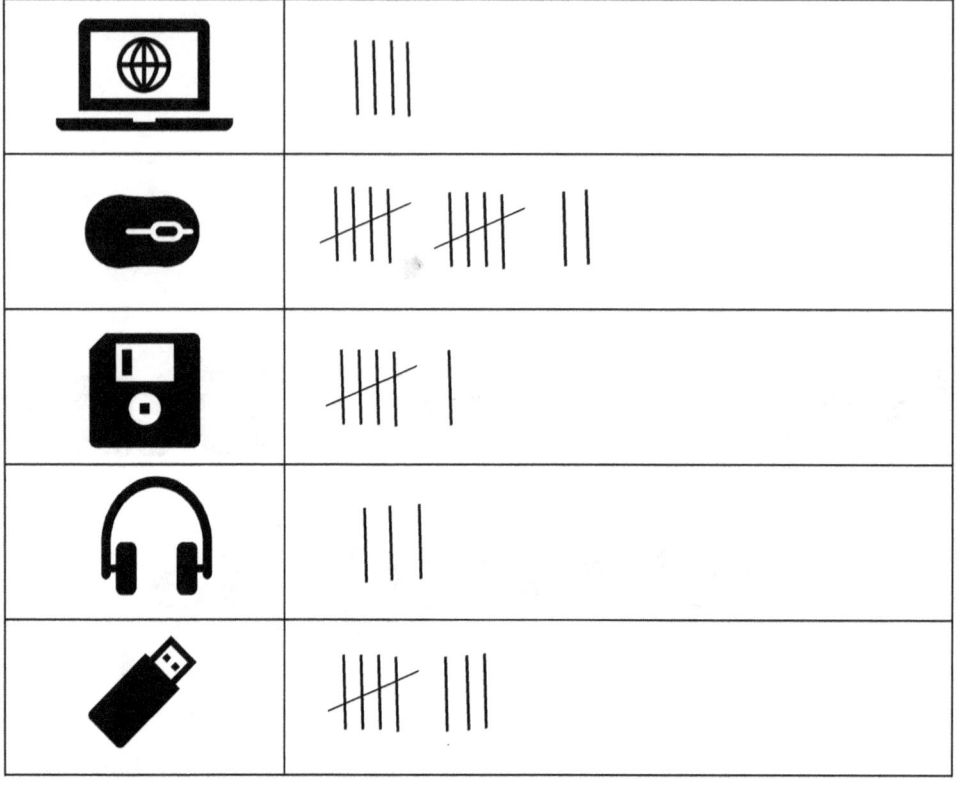

Key: ✦ = 2 Hardware

# Dot plots

The ages of students in a Math class are given below.

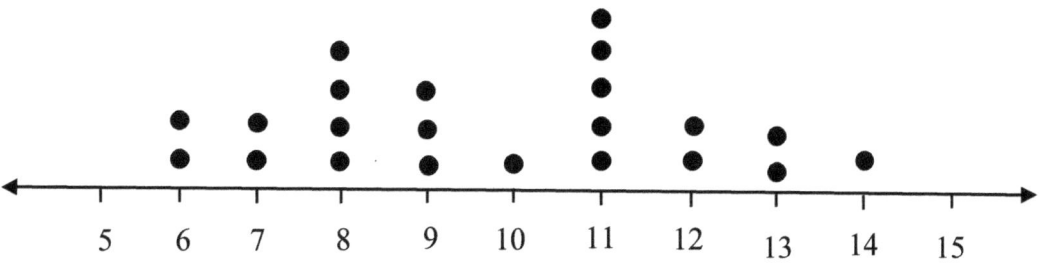

1) What is the total number of students in math class?

2) How many students are at least 12 years old?

3) Which age(s) has the most students?

4) Which age(s) has the fewest student?

5) Determine the median of the data.

6) Determine the range of the data.

7) Determine the mode of the data.

# Bar Graph

Each student in class selected two games that they would like to play. Graph the given information as a bar graph and answer the questions below:

| Game | Votes |
|---|---|
| Football | 13 |
| Volleyball | 10 |
| Basketball | 18 |
| Baseball | 17 |
| Tennis | 13 |

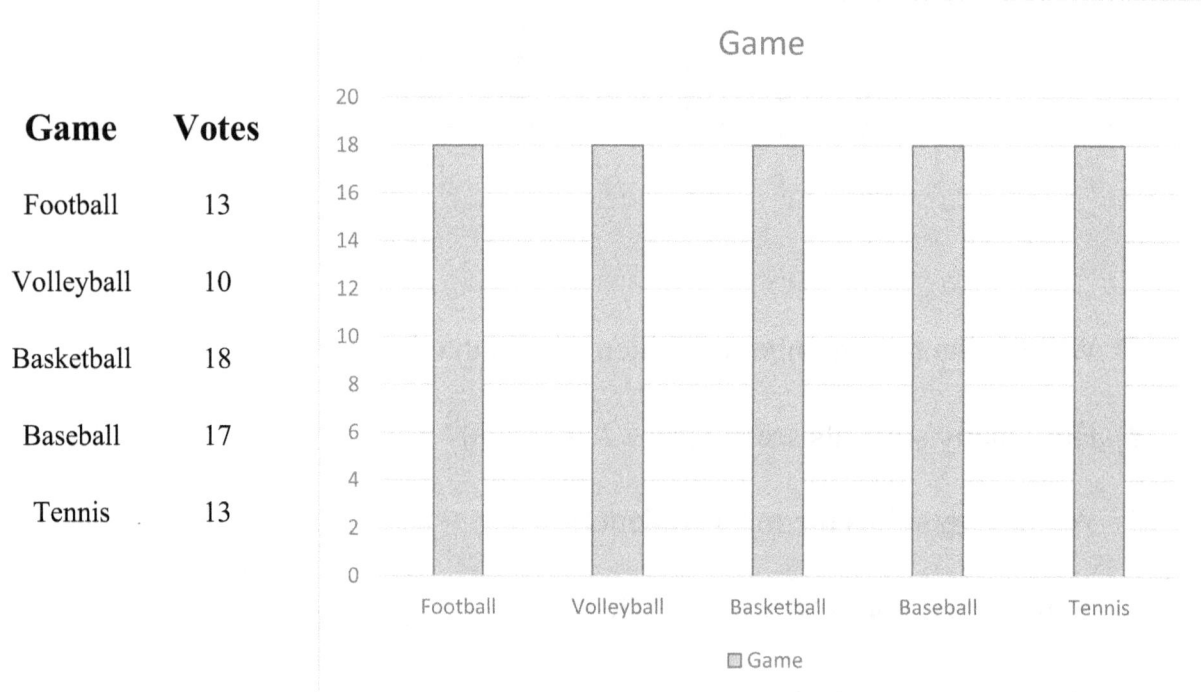

1) Which was the most popular game to play?

2) How many more students like Basketball than Volleyball?

3) Which two game got the same number of votes?

4) How many Volleyball and Football did student vote in all?

5) Did more student like football or Volleyball?

6) Which game did the fewest student like?

# Line Graphs

Amelia works in a doll store. She records the number of dolls sold in five days on a line graph. Use the graph to answer the questions.

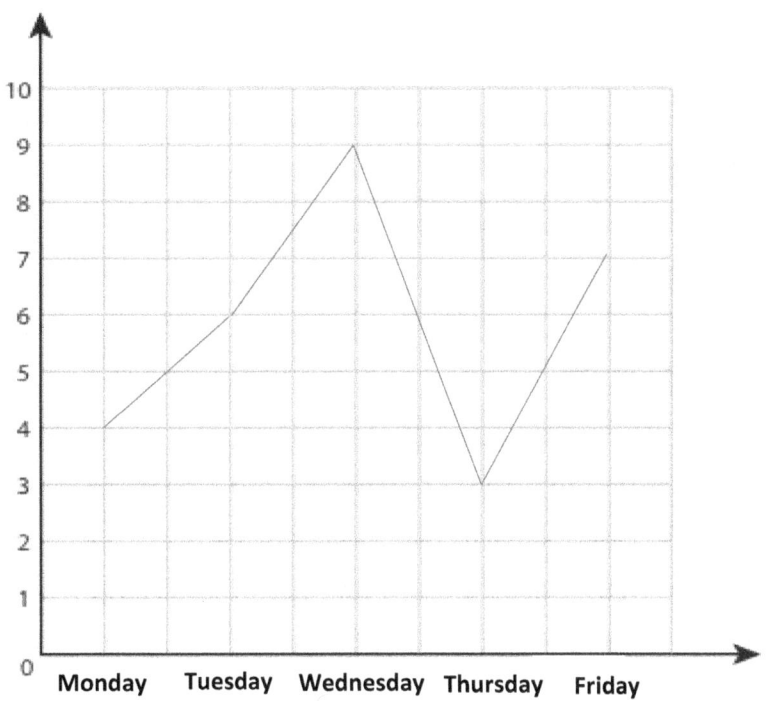

1) How many dolls were sold on Tuesday?

2) Which day had the minimum sales of dolls?

3) Which day had the maximum number of dolls sold?

4) How many dolls were sold in 5 days?

# Answer key Chapter 10

## Tally and Pictographs

| | |
|---|---|
| 💻 | ✳ ✳ |
| 🖱 | ✳ ✳ ✳ ✳ ✳ |
| 💾 | ✳ ✳ ✳ |
| 🎧 | ✳ ✳ |
| 🔌 | ✳ ✳ ✳ ✳ |

## Dot plots

1) 22
2) 5
3) 11
4) 10 and 14
5) 2
6) 4
7) 2

## Bar Graph

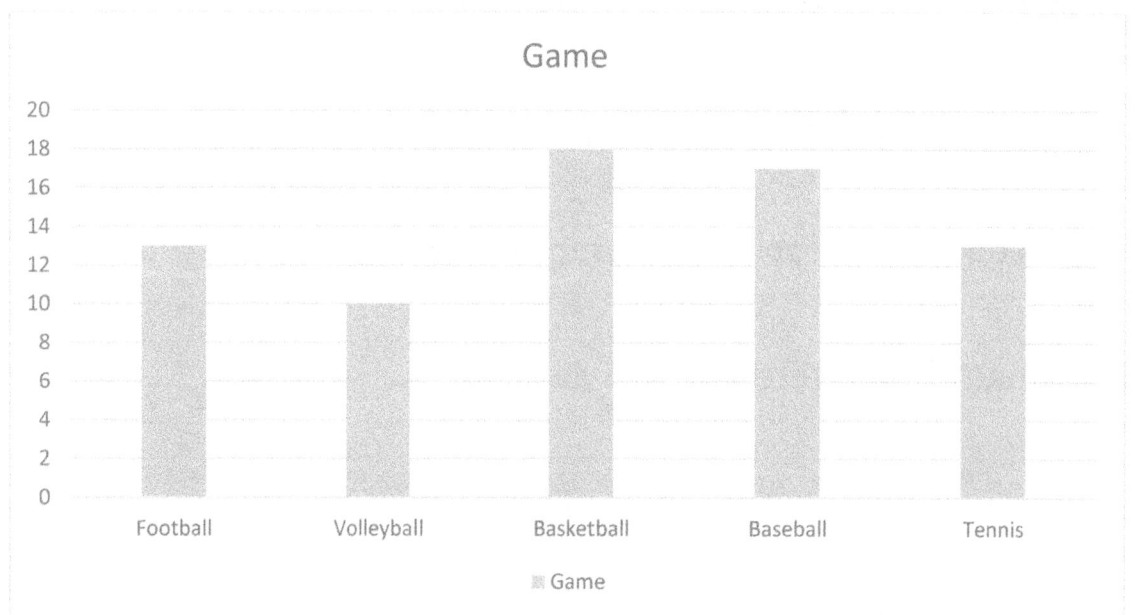

# STAAR Math Practice Grade 3

1) Basketball    3) Football and Tennis    5) Football
2) 8 students    4) 23    6) Volleyball

## Line Graphs

1) 6    2) Thursday    3) Wednesday    4) 29

# STAAR
# Test Review

## STAAR GRADE 3 MAHEMATICS REFRENCE MATERIALS

### LENGTH

| Customary | Metric |
|---|---|
| 1 mile (mi) = 1,760 yards (yd) | 1 kilometer (km) = 1,000 meters (m) |
| 1 yard (yd) = 3 feet (ft) | 1 meter (m) = 100 centimeters (cm) |
| 1 foot (ft) = 12 inches (in.) | 1 centimeter (cm) = 10 millimeters (mm) |

### VOLUME AND CAPACITY

| Customary | Metric |
|---|---|
| 1 gallon (gal) = 4 quarts (qt) | 1 liter (L) = 1,000 milliliters (mL) |
| 1 quart (qt) = 2 pints (pt.) | |
| 1 pint (pt.) = 2 cups (c) | |
| 1 cup (c) = 8 fluid ounces (Fl oz) | |

### WEIGHT AND MASS

| Customary | Metric |
|---|---|
| 1 ton (T) = 2,000 pounds (lb.) | 1 kilogram (kg) = 1,000 grams (g) |
| 1 pound (lb.) = 16 ounces (oz) | 1 gram (g) = 1,000 milligrams (mg) |

### Time

1 year = 12 months

1 year = 52 weeks

1 week = 7 days

1 day = 24 hours

1 hour = 60 minutes

1 minute = 60 seconds

# State of Texas Assessments of Academic Readiness

# STAAR Practice Test 1

## Mathematics

## GRADE 3

Administered *Month Year*

# STAAR Math Practice Grade 3

1) Which equation is true when the missing number is the number 9?

   A. $7 \times \underline{\phantom{?}?\phantom{?}} = 65$

   B. $14 \times \underline{\phantom{?}?\phantom{?}} = 126$

   C. $7 \times \underline{\phantom{?}?\phantom{?}} = 68$

   D. $14 \times \underline{\phantom{?}?\phantom{?}} = 116$

2) Which two fractions should be plotted at the same locations on a number line?

   A. $\frac{2}{7}$ and $\frac{6}{14}$

   B. $\frac{2}{7}$ and $\frac{8}{21}$

   C. $\frac{3}{7}$ and $\frac{15}{35}$

   D. $\frac{3}{7}$ and $\frac{9}{28}$

3) The table shows the number of visitors at the state parks over four months. Which comparison of these months is true?

   A. The visitors of April > The visitors of Jun

   B. The visitors of April < The visitors of May

   C. The visitors of May > The visitors of July

   D. The visitors of May < The visitors of June

| Month | Number of Visitors |
|---|---|
| April | 14,027 |
| May | 13,859 |
| Jun | 14,256 |
| July | 14,046 |

## STAAR Math Practice Grade 3

4) The expanded form of a number is 70,000 + 200 + 9. What is the standard form of the number?

   A. 7,209

   B. 72,009

   C. 70,029

   D. 70,209

5) Another way to write $5 \times 6$ is:

   A. $6 \times 6 \times 6 \times 6 \times 6$

   B. $\frac{1}{6} + \frac{1}{6} + \frac{1}{6} + \frac{1}{6} + \frac{1}{6}$

   C. $5 + 5 + 5 + 5 + 5 + 5$

   D. $5 + 6 + 5 + 6$

6) Ethan's goal is to save $182 to purchase his favorite Lego in three months.

   - In the first month, he saved $62.

   - In the second month, he saved $27.

   How much money does Ethan need to save in third month to be able to purchase his favorite Lego?

   A. $93

   B. $88

   C. $182

   D. $128

7) Javier wants to build a rectangular garden in his backyard. How much fencing he will need to buy if his garden measures 5feet by 8 feet?

   A. 22 feet

   B. 26 feet

   C. 16 feet

   D. 30 feet

8) What fraction of the shape is shaded?

   A. $\frac{3}{4}$

   B. $\frac{1}{2}$

   C. $\frac{2}{8}$

   D. $\frac{5}{8}$

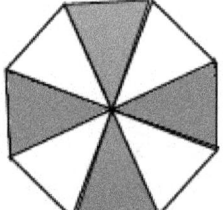

9) Andres drove his car 8,759 miles the first year. The second year, he drove 654 miles more than first year. What is the total number of miles he drove for two years?

   A. 9,413

   B. 9,341

   C. 18,172

   D. 18,372

10) Fatima asks people what their favorite hobbies are. She records their answers on the following bar graph. How many more people like traveling than gardening?

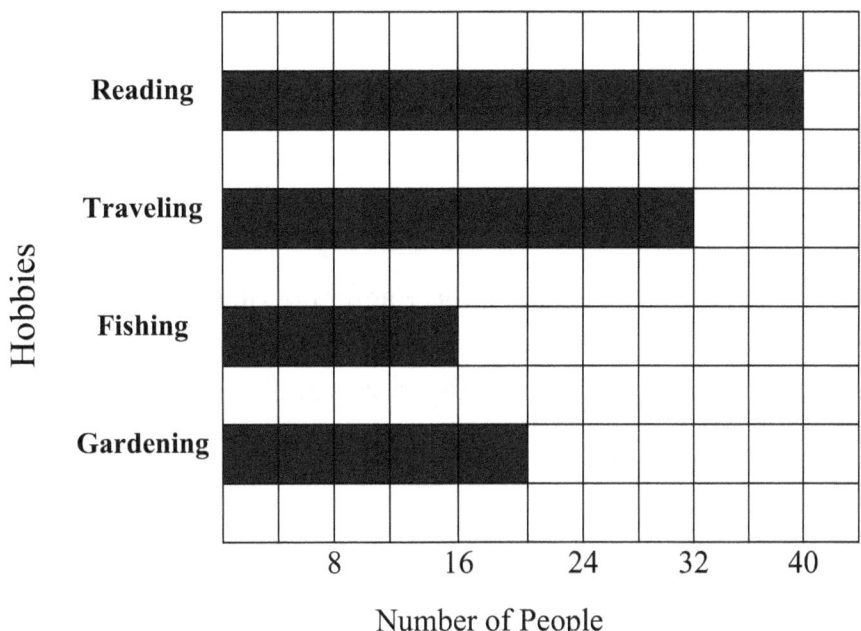

A. 6

B. 10

C. 12

D. 14

11) Time is on the P.M. clock. What time was it 3 hours and 20 minutes ago?

A. 10:30 A.M.

B. 11:30 P.M.

C. 9:30 P.M.

D. 11:45 A.M.

12) Mila counted her Lola toys. She put them in 2 groups of six and has 5 toys left over. How many toys does she have?

A. 10

B. 11

C. 17

D. 16

13) Erin bought three chocolate parfaits. Each chocolate parfaits cost 3 quarters, 2 nickels and 5 pennies. how much did she spend in all?

A. $2.25

B. $2.70

C. 90¢

D. 79¢

14) Two rectangles overlap, forming a smaller rectangle, as shown below. What is the area of shaded part of the figure shown?

A. 56 square meters

B. 96 square meters

C. 112 square meters

D. 104 square meters

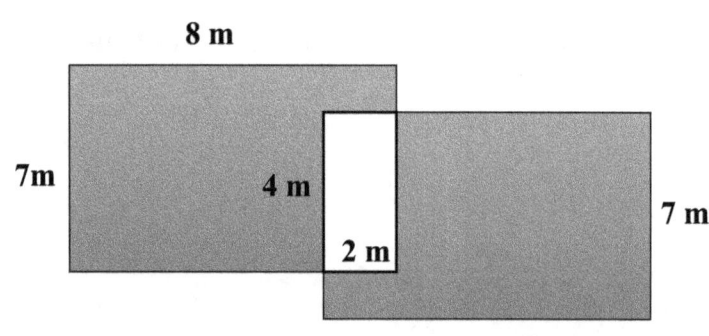

15) Shui baked 91 scones. If each scone tray held 7 scones, how many trays did Shui use?

A. 13

B. 23

C. 80

D. 424

16) Ms. Diaz ordered 6 pack of 8 markers. After passing out 2 markers to each student in her class, she has 6 left. How many students are in Ms. Diaz's class?

A. 21

B. 18

C. 42

D. 14

17) There are 45 slinkies star shape and 27 slinkies rainbow shape at a party. 9 children are given an equal number of each slinky. How many slinkies star and rainbow shape does each child get?

A. 6

B. 8

C. 18

D. 86

18) Pietro goes fishing at 10:28 a.m. he fishes for 45 minutes. What time is Pietro done fishing?

   A. 12:11 a.m.

   B. 12:13 a.m.

   C. 1:13 p.m.

   D. 11:13 a.m.

19) A box of 9 chocolates weighs 81 grams. If the empty box weighs 9 grams, how much does each chocolate weigh?

   A. 8

   B. 12

   C. 10

   D. 810

20) Look at the spinner above. On which toy is the spinner most likely to land?

   A. Dolls

   B. Car Toys

   C. Toy Animals

   D. All toys are equally likely

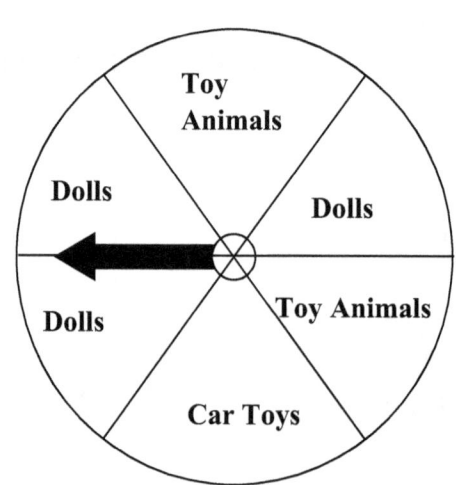

21) Which statement is correct?

   A. Even Number + Even Number = Odd Number

   B. Even Number + Odd Number = Even Number

   C. Odd Number + Odd Number = Odd Number

   D. Odd Number + Even Number = Odd Number

22) Kevin is driving 3 days to visit his parents. They lived 760 miles from Kevin. The first day he drove 112 miles. The second day he drove 184 miles. Which strip diagram can be used to find how many miles does he has to drive on the third day?

A.

| ? | | |
|---|---|---|
| 760 | 112 | 184 |

B.

| 760 | | |
|---|---|---|
| 112 | 184 | ? |

C.

| 184 | | |
|---|---|---|
| 760 | 112 | ? |

D.

| 112 | | |
|---|---|---|
| 760 | 184 | ? |

23) Ms. Evalyn asked her students which fruit they liked the most. The list shows the fruits they preferred.

16 Strawberries          24 bananas

12 cherries              32 pears

Which pictograph best represents the information in the list?

A.

| Favorite Fruits | |
| --- | --- |
| Strawberry | ✺ ✺ |
| cherry | ✺ ✺ ˅ |
| Banana | ✺ ✺ ✺ ✺ |
| Pear | ✺ ✺ ✺ ✺ ✺ |
| Each means 8 fruits | |

B.

| Favorite Fruits | |
| --- | --- |
| Strawberry | ✺ ✺ |
| cherry | ✺ ˃ |
| Banana | ✺ ✺ |
| Pear | ✺ ✺ ✺ |
| Each means 12 fruits | |

C.

| Favorite Fruits | |
| --- | --- |
| Strawberry | ✺ ˃ |
| cherry | ✺ ✺ ˅ |
| Banana | ✺ ✺ ˃ |
| Pear | ✺ ✺ ✺ |
| Each means 16 fruits | |

D.

| Favorite Fruits | |
| --- | --- |
| Strawberry | ✺ |
| cherry | ˃ |
| Banana | ✺ ˃ |
| Pear | ✺ ✺ |
| Each means 16 fruits | |

# STAAR Math Practice Grade 3

24) Mei and Lei can read at the same speed. If Mei can read 12 books every 96 days, how many books can Lei read in 120 days?

   A. 15

   B. 45

   C. 42

   D. 82

25) Which fraction is equivalent to 7?

   A. $\frac{7}{7}$

   B. $\frac{28}{4}$

   C. $\frac{1}{7}$

   D. $\frac{4}{7}$

26) Nora bought 5 boxes of cookies. Each box had 8 cookies in it. If ☐ represents the total number of cookies she bought, which equation can be used to find the total number of cookies Nora bought?

   A. $5 \times \square = 8$

   B. $\square \times 8 = 5$

   C. $5 + 8 = \square$

   D. $\square \div 5 = 8$

27) Which story best fits the equation $35 \div 7 = 5$ ?

A. Harry enter a peach-eating contest. He needs to eat 35 peaches to win. If he has eaten 7 peaches so far, how many peaches does he still have to eat?

B. James eats 7 peaches every day. How many peaches will he eat in 35 days?

C. Liam and his 6 friends ate a total of 35 peaches. If they each ate the same number of peaches, how many did they each eat?

D. A cartoon contains 35 peaches. If Oliver bought 7 cartoons, how many peaches did he buy?

28) The diameter of a circle is 16 feet. What is the length of its radius?

A. 8 feet

B. 16 feet

C. 32 feet

D. 64 feet

29) Willie bought 4 mechanical pencils for $12 each and 7 paper notebooks for $8 each. How much change did he get from $120?

A. $86

B. $104

C. $26

D. $16

30) The perimeter of regular hexagon below is 54. What is the length of MN?

A. 8

B. 9

C. 10

D. 18

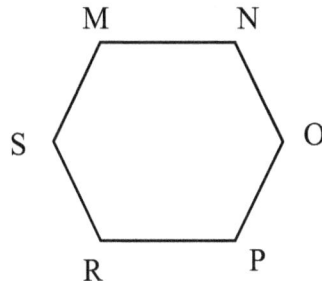

# State of Texas Assessments of Academic Readiness

# STAAR Practice Test 2

Mathematics

GRADE 3

Administered Month Year

# STAAR Math Practice Grade 3

1) How many of the following numbers are odd?

   19, 24, 53, 80, 75, and 21

   A. 2

   B. 4

   C. 5

   D. 6

2) What is this number: 7hundreds, 3 ones, 9thousands?

   A. 9,703

   B. 739,000

   C. 9573,000

   D. 3,709,000

3) Adela spent her time in below activities. What was the total amount of time she spent on all the activities?

   - She spent 27 minutes practicing guitar.
   - She spent 19 minutes playing game.
   - She spent 52 minutes jogging with her friend.

   A. 38 minutes

   B. 1 hour 8 minutes

   C. 1 hour 28 minutes

   D. 1 hour 38 minutes

## STAAR Math Practice Grade 3

4) Adriano bought 2 books of poetry cost $13.70. The other book cost $18.65.

   How much more did the second book of poetry cost than the first one?

   A. $6.15

   B. $4.95

   C. $14.50

   D. $26.95

5) What fraction of these numbers are greater than or equal to 59?

   12, 15, 27, 29, 46, 72, 40, 19, 54, 71, 60, 53, 28, 39, 58, 89

   A. $\frac{3}{8}$

   B. $\frac{3}{4}$

   C. $\frac{1}{4}$

   D. $\frac{3}{16}$

6) Ophelia has 126 guests coming to her party.

   - Each table will hold 7 guests.
   - Each table has 3 silk roses.

   How many silk roses will she need?

   A. 18

   B. 54

   C. 36

   D. 114

7) Round 193 to the nearest tens?

   A. 203

   B. 190

   C. 200

   D. 1,930

8) Calla drew a very large rectangle with a red piece of chalk at the playground. The length is 6 feet and the width is 4 feet. Calla can only walk on the line. If she wants to walk the rectangle 2 times by only stepping on the line, how many feet will he end up walking?

   A. 20

   B. 45

   C. 40

   D. 24

9) Aiden biked 4 miles last week. Adrian biked 36 miles last week. Adrian biked how many times as many miles as Aiden? Which equation can help you answer the question?

   A. $4 \times 36 = \Box$

   B. $4 + 36 = \Box$

   C. $36 - 4 = \Box$

   D. $36 \div 4 = \Box$

10) A clothing store is having a sale on shirts. The new price of each shirt is $6 less than the old price. Which table shows prices of different shirts at this store?

A.

| Shirt Sale | | | | |
|---|---|---|---|---|
| Old Price | $7 | $14 | $24 | $43 |
| New Price | $11 | $20 | $30 | $49 |

B.

| Shirt Sale | | | | |
|---|---|---|---|---|
| Old Price | $4 | $6 | $12 | $19 |
| New Price | $10 | $12 | $16 | $24 |

C.

| Shirt Sale | | | | |
|---|---|---|---|---|
| Old Price | $12 | $18 | $28 | $39 |
| New Price | $6 | $14 | $20 | $33 |

D.

| Shirt Sale | | | | |
|---|---|---|---|---|
| Old Price | $10 | $12 | $18 | $24 |
| New Price | $4 | $6 | $12 | $19 |

11) Ashley is putting together goodie bags for her birthday party. She invited 10 friends, and everyone can come except for Lisa. At the party store, she bought 37 flower hair clips. She wants to give everyone an equal number of hair clips. How many should she put into each goodie bag?

A. 9

B. 4

C. 1

D. 5

12) Select the number that is the largest?

   A. 30,033

   B. 30,300

   C. 30,030

   D. 30,003

13) Brady learned 7 new spelling words last week. Dylan learned 6 times as many words as Brady. How many words did Dylan learn? Draw a bar model to find the number of words Dylan learned.

   A.

   | 6 | 6 | 6 | 7 | 7 | 7 |

   ?

   B.

   | 7 | 7 | 7 | 7 | 7 | 7 |

   ?

   C.

   | 6 | 6 | 6 | 6 | 6 | 6 |

   ?

   D.

   | 6 | 7 |

   ?

14) Adam held out his hand. He had 1 half dollar, 2 quarter, 3 dimes, and 1 pennies in it. What amount did he have in his hand?

   A. $1.31

   B. $131

   C. $0.131

   D. $1.13

15) There are 5 pieces of cranberry walnut fudge on each package. There are 18 packages. How many pieces of cranberry walnut fudge are there in all?

   A. 92

   B. 88

   C. 80

   D. 90

16) Ming made roasted garlic soup for the appetizer. She made 14 cups of the soup. At the end of the lunch there were $5\frac{1}{3}$ cups of soup left. How many cups of soup were eaten at the lunch?

   A. $14\frac{1}{3}$

   B. $8\frac{1}{3}$

   C. $10\frac{1}{3}$

   D. $8\frac{2}{3}$

17) A school uses 4 school buses to take student on a field trip. There are 32 students on each bus. How many students are on the field trip?

A. 28

B. 36

C. 132

D. 128

18) A mailman brought our mail at 10:13 a.m. today. Yesterday he brought it at 2:48 p.m. how many minutes later was the mailman yesterday?

A. 4 hours and 35 minutes

B. 4 hours and 25 minutes

C. 3 hours and 45 minutes

D. 2 hours and 35 minutes

19) Mila has a dark chocolate bar. She breaks it into fourth. She gave $\frac{1}{3}$ of the chocolate bar to her friend. Which expression represents the fraction of chocolate bar Mila has left?

A. $\frac{1}{9} + \frac{1}{9}$

B. $\frac{1}{2} + \frac{1}{2}$

C. $\frac{1}{3} + \frac{1}{3}$

D. $\frac{1}{4} + \frac{1}{4}$

20) What is the area, in square meters, of the shape below?

A. 40 square meters

B. 48 square meters

C. 88 square meters

D. 120 square meters

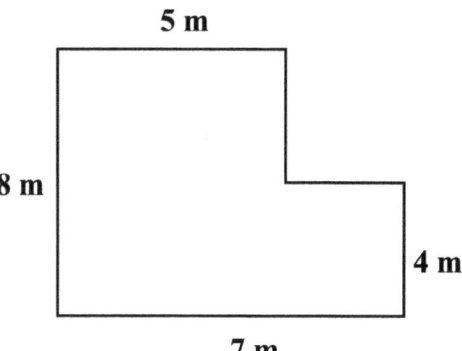

21) Anna's fish tank holds 54 liters of water. He uses a 6-liters bucket to fill the tank. How many buckets of water are needed to fill the tank?

A. 6

B. 9

C. 15

D. 60

22) Which number line shows the correct locations of all given values?

$$\frac{4}{5}, \frac{13}{10}$$

A.

B.

C.

D.

23) Last week, Jason's website had 4,251 visitors and Evan's website had 867 fewer visitors than Jason. Peter's website had 633 fewer visitors than Evan. How many visitors did Peter's website last week?

A. 2,751

B. 2,517

C. 4,017

D. 5,352

24) The sum of a two-digit number and on-digit number is 59. The product of the two numbers is 364. What are the two numbers?

A. 91 and 4

B. 52 and 7

C. 53 and 6

D. 50 and 9

25) A shape is divided into equal parts as shown. Which two equivalent fractions can represent the unshaded area of shape?

A. $\frac{1}{4}$ and $\frac{2}{6}$

B. $\frac{3}{8}$ and $\frac{3}{6}$

C. $\frac{2}{3}$ and $\frac{4}{6}$

D. $\frac{2}{3}$ and $\frac{3}{6}$

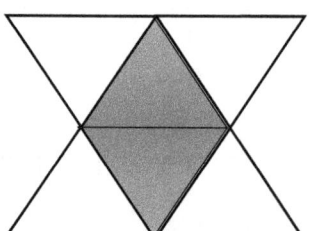

26) A school collected following data from some students about how far they live from the school. Which dot plot represented the distance in mile of each student?

2, 5, 2, 3, 2, 5, 3, 4, 3, 1, 4, 2

A.
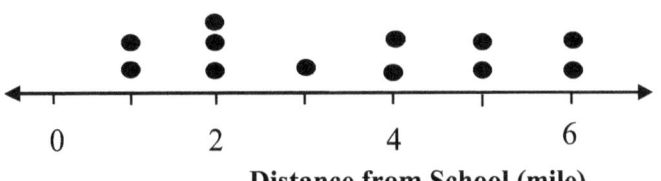

B.
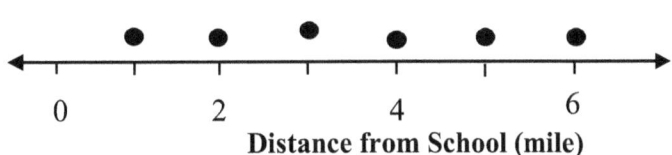

C.
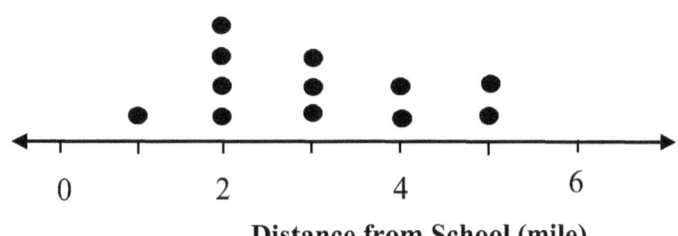

D.
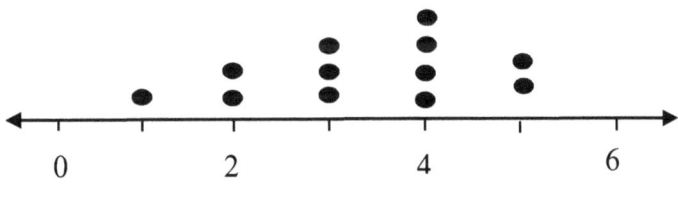

27) Which expression is equivalent to $(8 + 2) \times 5$ ?

A. $(5 \times 8) + (5 \times 2)$

B. $(8 \times 5) + (8 \times 2)$

C. $2 \times (5 \times 8)$

D. $(8 + 5) \times (2 + 5)$

28) The floor of a yard kennel is covered by square tiles as shown. What is the area, in square feet, of the yard kennel? ( = 1 square foot)

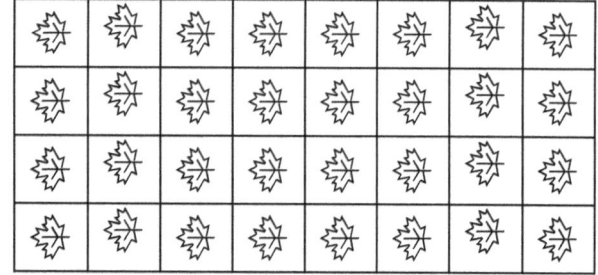

A. 18 square feet

B. 16 square feet

C. 42 square feet

D. 32 square feet

29) Last year there were 358 students at Marvell Montessori. This year, 213 more students enrolled. How many students attendant Marvell Montessori?

A. 145

B. 571

C. 472

D. 853

30) Itzel had 32 crayons on her table. She buys new boxes of crayon that have 7 crayons in each box. Now, she has 95 crayons. how many new boxes did he buy?

A. 6

B. 9

C. 11

D. 15

# Answers and Explanations

# STAAR Math Practice Grade 3

# Answer Key

Now, it's time to review your results to see where you went wrong and what areas you need to improve!

## STAAR Math Practice Tests

### Practice Test 1

| | | | | | |
|---|---|---|---|---|---|
| 1 | B | 11 | A | 21 | D |
| 2 | C | 12 | C | 22 | B |
| 3 | D | 13 | B | 23 | D |
| 4 | D | 14 | B | 24 | A |
| 5 | C | 15 | A | 25 | B |
| 6 | A | 16 | A | 26 | D |
| 7 | B | 17 | B | 27 | C |
| 8 | B | 18 | D | 28 | A |
| 9 | C | 19 | A | 29 | D |
| 10 | C | 20 | A | 30 | B |

### Practice Test 2

| | | | | | |
|---|---|---|---|---|---|
| 1 | B | 11 | B | 21 | B |
| 2 | A | 12 | B | 22 | C |
| 3 | D | 13 | B | 23 | A |
| 4 | B | 14 | A | 24 | B |
| 5 | C | 15 | D | 25 | C |
| 6 | B | 16 | C | 26 | C |
| 7 | B | 17 | D | 27 | A |
| 8 | C | 18 | A | 28 | D |
| 9 | D | 19 | C | 29 | B |
| 10 | D | 20 | B | 30 | B |

# Practice Test 1

# Answers and Explanations

**1) Answer: B**

$14 \times 9 = 126$

**2) Answer: C**

Equivalent fractions have different names but the same value; they are at the same place on the number line. To find equivalent fractions, multiply the numerator AND denominator by the same nonzero whole number.

**3) Answer: D**

May has the least visitors. April, July, and June are other months in order from least to greatest.

**4) Answer: D**

$70,000 + 200 + 9 = 70,209$

**5) Answer: C**

The multiplication of two whole numbers, when thinking of multiplication as repeated addition, is equivalent to adding as many copies of one of them (multiplicand) as the value of the other one (multiplier).

$5 \times 6 = 6 + 6 + 6 + 6 + 6 = 5 + 5 + 5 + 5 + 5 + 5$

**6) Answer: A**

Ethan saved $62 and $27. Therefore, he has $87 now. $62 + $27 = $89, $182 − $89 = $93. He needs to save 93.

**7) Answer: B**

Use perimeter of rectangle formula.

Perimeter = 2 × length + 2 × width ⇒ P= (2 × 5) + (2 × 8) = 10 +16 = 26 feet

**8) Answer: B**

A fraction is the number of shaded parts divided by the number of equal parts:

number of shaded parts: numerator

number of equal parts: denominator

then, the answer is: $\frac{4}{8} = \frac{1}{2}$

## 9) Answer: C

The second year: $8{,}759 + 654 = 9{,}413$ miles

Total number of miles = number of miles the first year + number of miles the second year

Total number of miles = $8{,}759 + 9{,}413 = 18{,}172$

## 10) Answer: C

Number of people like traveling: 32

Number of people like gardening: 20

$32 - 20 = 12$ more people

## 11) Answer: A

The clock shows 1:50 PM. 3 hours before that was 10:50 AM. 20 minutes before that was 10:30 AM.

## 12) Answer: C

$2 \times 6 = 12$, then add the remained toys: $12 + 5 = 17$

## 13) Answer: B

We have added the values of the 3 quarters, 2 nickels, and 5 pennies:

$25 + 25 + 25 + 5 + 5 + 1 + 1 + 1 + 1 + 1 = 90$ cents cost of each chocolate parfaits.

$90 + 90 + 90 = 270$ cents = $2.70

## 14) Answer: B

Area of shaded part = area of two rectangle − area of two overlapped rectangle

Area of rectangle = length × width = $8 \times 7 = 56$ cm

Area of overlapped rectangle = $4 \times 2 = 8$ cm

Area of shaded part = $(2 \times 56) + (2 \times 8) = 112 - 16 = 96$ cm

## 15) Answer: A

Shui wants to put 91 scones into tray of 6 scones.

Therefore, she needs $91 \div 7 = 13$ trays.

## 16) Answer: A

$6 \times 8 = 48$ markers, she passed out 42 ($48 - 6 = 42$)

Each student has 2 markers, then $42 \div 2 = 21$ students

17) **Answer: B**

Number of slinky star shape: $45 \div 9 = 5$

Number of slinky rainbow shape: $27 \div 9 = 3$

Each child gets: $5 + 3 = 8$

18) **Answer: D**

28 min + 45 min = 73 min = 1:13; 10 a.m. + 1:13 = 11:13 a.m.

19) **Answer: A**

Subtract 9 from 81: $81 - 9 = 72$; $72 \div 9 = 8$

20) **Answer: A**

The chance of landing on toy animals is 2 out of 6.

The chance of landing on car toys is 1 out of 6.

The chance of landing on dolls is 3 out of 6.

The chance of landing on dolls is more than the chance of landing on other toys.

21) **Answer: D**

Let us check the options provided by substituted the odd and even numbers:

A. Even Number + Even Number = Odd Number → $2 + 4 = 6$ even, is not correct.

B. Even Number + Odd Number = Even Number → $2 + 5 = 7$ odd, is not correct.

C. Odd Number + Odd Number = Odd Number → $3 + 5 = 8$ even, is not correct.

D. Odd Number + Even Number = Odd Number → $5 + 2 = 7$ odd, is correct!

22) **Answer: B**

You should have analyzed each model to determine which one shows that when added, the 112 miles in the first day, plus the 184 miles in the second day, plus the miles in the third day are equal to the total number of miles (760).

23) **Answer: D**

You should have multiplied the number of whole icons shown in each row by the number shown in the key and the number of fraction icons ($\frac{1}{2}, \frac{1}{4}$) by fraction the number shown in the key.

You found that there are 16 strawberries 12 cherries 24 banana and 32 pears as listed in the problem. (key number is 16, half of 16 is 8, $\frac{1}{4}$ of 16 is 4)

**24) Answer: A**

First find the speed of Mei: $96 \div 12 = 8$

then, 120 days divided by the same speed (8): $120 \div 8 = 15$

**25) Answer: B**

To find equivalent fractions, you should have written any integer number as fraction number ($\frac{7}{1}$) then multiply the numerator AND denominator by the same.

$$\frac{7}{1} \times \frac{4}{4} = \frac{28}{4}$$

**26) Answer: D**

When we divide, we look to separate into equal groups, while multiplication involves joining equal groups. Multiplication and division are closely related, given that division is the inverse operation of multiplication.

If we have $5 \times 8 = \Box$, its inverse relations (in the form of a division) will be:

$\Box \div 5 = 8$ or $\Box \div 8 = 5$

**27) Answer: C**

To determine which statement represents, you should count total number of something (peaches) equally distributed (divided) into same bundles (persons).

**28) Answer: A**

The diameter is the distance from one side of the circle to the other at its widest points. The diameter will always pass through the center of the circle. The radius is half of this distance.

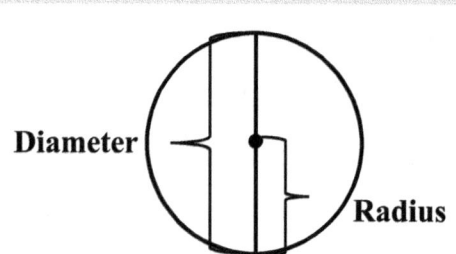

$16 \div 2 = 8$

**29) Answer: D**

$4 \times \$12 = \$48$, and $7 \times \$8 = \$56$

$\$120 - (\$48 + \$56) = \$120 - \$104 = \$16$

**30) Answer: B**

When all angles are equal and all sides are equal it is regular, otherwise it is irregular.

A hexagon is a 6-sided polygon. Then divided the perimeter by 6 to find the length of each side: $54 \div 6 = 9$

# Practice Test 2
## Answers and Explanations

**1) Answer: B**

Odd numbers have the digits 1, 3, 5, 7 or 9 in their ones place.

19, 53,75, and 21 are odd.

**2) Answer: A**

9thousands, 7hundreds, 3 ones equal 9,703

**3) Answer: D**

$27 + 19 + 52 = 98$ minutes, that 60 minutes is equal to 1 hour, 98 minutes is equal to 1 hour 38 minutes.

**4) Answer: B**

We need to subtract the cost of two books: $18.65 - 13.70 = 4.95$

**5) Answer: C**

A fraction represents a part of a whole. Order the numbers from least to greatest:

12, 15, 19, 27, 28, 29, 39, 40, 46, 53, 54, 58, 60, 71, 72, 89

Fraction $= \frac{part}{whole} = \frac{4}{16}$ and equivalent fraction is: $\frac{1}{4}$

**6) Answer: B**

Find the number of tables: $126 \div 7 = 18$

Silk rose: $18 \times 3 = 54$

**7) Answer: B**

Rule for rounding to the nearest tens: Look at the number in the one's place and...

For 0, 1, 2, 3 or 4, we round down

For 5, 6, 7, 8 or 9, we round up

193 rounds to 190

**8) Answer: C**

Perimeter is the distance around a shape. The perimeter of a rectangle is the total length of all the sides of the rectangle. you can find the perimeter by adding all four sides of a rectangle.

$6 + 4 + 6 + 4 = 20$, then 2 times is $2 \times 20 = 40$ feet

## 9) Answer: D

We may have divided 36 by 4, or thought, how many times as many as 4 is 36?

## 10) Answer: D

You should have subtracted $8 from each regular price of a shirt and used the result to confirm each sale price listed in the table. New price (Sale price) = Old price $-6$

$10 - 6 = 4, 12 - 6 = 6, 18 - 6 = 12, 24 - 6 = 19$

## 11) Answer: B

She invited 10 friends and one of them cannot come: $10 - 1 = 9$ friends,

37 flower hair clips divided by 9 friends: $37 \div 9 = 4 \: R \: 1$

4 numbers in each goodie bag and 1 number left

## 12) Answer: B

| Place Value Chart | | | | | |
|---|---|---|---|---|---|
| Thousands | | | Ones | | |
| Hundreds | Tens | Ones | Hundreds | Tens | Ones |
| | First | Second | Third | Fourth | Fifth |

when 5-digit numbers are compared start with ten thousands place (First) and then one thousands place (Second), etc. If one whole number has a higher number in the tens thousands place, then it is larger than a whole number with fewer ten thousands. If the ten thousands are equal compare the one thousands, then the hundreds (Third), tens (Fourth), and ones (Fifth). when compared the numbers provided; ten thousand and one thousand places are the same. 3 in hundreds place is higher than other hundreds places.

## 13) Answer: B

The strip diagram should use same-size sections to model the multiplication problem $6 \times 7 =?$, where ? represents the total words Dylan learned. The strip diagram shows 6 same size sections representing the words learned with 7 in each section.

## 14) Answer: A

We could have added the values of the 1 half dollar, 2 quarter, 3 dimes, and 1 pennies:

$50 + 25 + 25 + 10 + 10 + 10 + 1 = 131$ cents and then changed to dollar notation:

131 cents = $1.31

**15) Answer: D**

We have multiplied the 5 pieces on each package by 18 packages: (5 × 18 = 90)

**16) Answer: C**

you can rewrite the whole number as a mixed number to perform the subtraction. You use an equivalent mixed number that has the same denominator as the fraction in the other mixed number ($14 = 13 + 1 = 13 + \frac{3}{3} = 13\frac{3}{3}$).

$14 - 5\frac{1}{3} = 13\frac{3}{3} - 5\frac{1}{3}$ then, $(13 - 5 = 8$, and $\frac{3}{3} - \frac{1}{3} = \frac{2}{3})$, add $8 + \frac{2}{3} = 8\frac{2}{3}$ cups

**17) Answer: D**

4 × 32 = 128 students

**18) Answer: A**

From 10:13 A.M. to 2:48 P.M.: 4 hours and 35 minutes

**19) Answer: C**

$1 - \frac{1}{3} = \frac{3}{3} - \frac{1}{3} = \frac{2}{3}$, and $\frac{1}{3} + \frac{1}{3} = \frac{2}{3}$

**20) Answer: B**

You may have split the figure into two shapes and found the areas of both shapes, and then added the areas of the two shapes together.

Shape 1 (rectangle): 8 m × 5 m = 40 square meters

Shape 2 (rectangle): $(7 - 5) m \times 4 m =$

$2 m \times 4 m = 8$ square meters

40 + 8 = 48 square meters

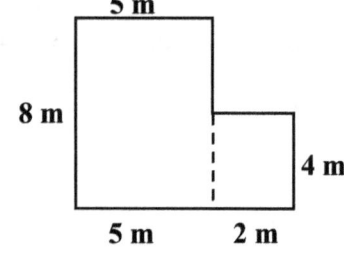

**21) Answer: B**

You could have divided 54 liters by 6-liters bucket: 54 ÷ 6 = 9

**22) Answer: C**

First counted the number of sections on the number line between two numbers, since there are 10 section between 0 and 1 or 1 and 2, each section represented $\frac{1}{10}$. Then for $\frac{4}{5} = \frac{8}{10}$, you count 8 sections after 0 and for $\frac{13}{10} = \frac{10}{10} + \frac{3}{10} = 1 + \frac{3}{10} = 1\frac{3}{10}$, you count 3 sections after 1.

## STAAR Math Practice Grade 3

**23) Answer: A**

Evan' website visitors: $4{,}251 - 867 = 3{,}384$

Peter's website visitors: $3{,}384 - 633 = 2{,}751$

**24) Answer: B**

If we review the options provided, then the option B is correct.

$52 \times 7 = 364$, and $52 + 7 = 59$

**25) Answer: C**

Write a fraction for the model. The number of **un**shaded parts is the numerator (top number) and the total number of parts is the denominator (bottom number) of the fraction. Then write an equivalent fraction for the model.

Model shows $\frac{4}{6}$, and the equivalent is $\frac{2}{3}$

**26) Answer: C**

To determine the correct dot plot (graph that uses dots to display data), we should have sorted the distances by value: 1, 2, 2, 2, 2, 3, 3, 3, 4, 4, 5, 5

And counted the number of distances for each value. You have chosen the dot plot with one dots above the value 1, four dots above the value 2, three dots above the value 3, two dots above the value 4 and two dot above the value 5.

**27) Answer: A**

The distributive property states that multiplying the sum of two or more addends by a number will give the same result as multiplying each addend individually by the number and then adding the products together. It helps in making difficult problems simpler.

**28) Answer: D**

Use area formula of a rectangle: Area = length × width

Area = 8 feet × 4 feet = 32 square feet

**29) Answer: B**

$358 + 213 = 571$

## 30) Answer: B

First count how many crayons did she buy: $95 - 32 = 63$

And find the boxes: $63 \div 7 = 9$

"End"

www.ingramcontent.com/pod-product-compliance
Lightning Source LLC
LaVergne TN
LVHW061311060426
835507LV00019B/2105